People, Parasites, and Plowshares

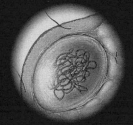

People, Parasites, and Plowshares

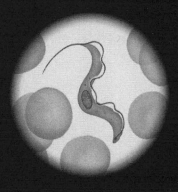

Learning from Our Body's Most Terrifying Invaders

DICKSON D. DESPOMMIER

Foreword by William C. Campbell

Columbia University Press
New York

Columbia University Press
Publishers Since 1893
New York   Chichester, West Sussex
cup.columbia.edu

Library of Congress Cataloging-in-Publication Data
Despommier, Dickson D.

People, parasites, and plowshares : learning from our body's most terrifying invaders /
Dickson D. Despommier ; foreword by William C. Campbell.

pages cm
Includes bibliographical references and index.
ISBN 978-0-231-16194-7 (cloth : alk. paper)—ISBN 978-0-231-16195-4 (pbk. : alk.
paper)—ISBN 978-0-231-53526-7 (ebook)

1. Parasites—Popular works.   2. Host-parasite relationships—Popular works.
3. Parasitology.   I. Title.

QL757.D47 2013
578.6'5—dc23

2012050589

COVER PHOTO: David Scharf © Getty Images
COVER DESIGN: Milenda Nan Ok Lee

CONTENTS

ILLUSTRATIONS

FOREWORD

O! your parasite
Is a most precious thing . . .
—BEN JONSON (1607)

No less a genius than Alfred North Whitehead was of the opinion that one is unlikely to have new ideas after the age of 60, but one can still make good use of the ideas one already has. It is no secret that Dr. Dickson Despommier's genius has been more than sixty years in the making, yet in his conversation new ideas burst forth like bullets from a machine gun. Many of his friends have experienced the phenomenon firsthand—an experience that his former colleague Dr. George Stewart has referred to as being "dicksonized." The book *People, Parasites, and Plowshares*, however, epitomizes the other half of Whitehead's assertion. Over the years, Despommier has written much that is addressed to professional colleagues in medicine and the biomedical sciences; now he aims to share his knowledge of parasitology with the general reader. His knowledge is vast and his aim is true. The result is a work that is eminently instructive as well as being fun for the reader and (quite evidently) for the author. But that is not the only reason that his writing of this book has been a worthwhile endeavor.

Other disciplines, besides parasitology, bring special techniques and insights to the study of parasites, and from them we learn subtle secrets that can be uncovered only by biochemistry, pathophysiology, immunology, and many other branches of science. So, by steps big and small, our understanding of parasitism is constantly enriched. Despommier teaches us the basic elements of medical parasitology (drawing on his long and much acclaimed tenure on the professorial podium) but also brings to the pages of this book a treasure of the latest advances in the most arcane and up-to-date areas of research. These advances he integrates artfully into his narrative along with bold and plentiful speculation as to their promise or lack of promise in practical affairs. Those of us who exceed the Whitehead boundary by an even wider margin may rub our eyes in astonishment, even bewilderment, but young readers will relish this feast of information—and many, we may hope, will be inspired by it. In this context it may be worth noting the value of introducing young students to medical parasitology for they are already exposed to modern scientific concepts and, besides, they tend to have a natural affinity for the grotty kind of visual image that medical parasitology produces in such profusion.

Though parasitology is not an endangered discipline, parasitologists sometimes worry that they are becoming an endangered species. This is because hordes (and I use the word respectfully) of scientists do research on parasites day in and day out, while a smaller and smaller proportion of them come up through the traditional avenues of training and confraternal bonding that mean so much to those of us who are "old-school" parasitologists. Such institutional evolution engenders a trace of foreboding—a worry that interest as well as expertise is fragmenting, thereby constricting the reach of parasitology as an academic fellowship. Eventually the expanding and splintering disciplines of science may collapse into a dense mass of physics; but that is hardly imminent. For now, a book like *People, Parasites, and Plowshares* is just the sort of tonic we need.

To this day, the traditional sort of parasitology often comes up with startling new findings. The worm *Onchocerca volvulus*, which causes river blindness, is itself parasitized by bacteria. (I like to imagine, just for fun, that Shakespeare's Prospero had Onchocerca in mind when he exclaimed, "Poor worm, thou art infected!") Only recently has it been recognized that the bacteria in the worms play a role in causing the actual signs and symptoms of the human disease. While the clinical manifestations are bad enough, the social consequences add greatly to the misery of affected populations. I have known about river blindness since I was a student and have seen its ravages in sub-Saharan Africa, yet never have I seen so graphic an account of its social impact as in the pages of this book. (Inclusion of such vignettes is one of the bonuses the book provides.) For well over a century, the tiny roundworm *Trichinella spiralis* has been known to cause trichinellosis (trichinosis) in humans and to live in the flesh of pigs. Traditional methods have also shown that the worm exists in many warm-blooded animals—and even in the cold-blooded crocodile—and that the Arctic can boast a Trichinella species that can survive for many months in frozen meat! Studies in molecular and evolutionary biology have elucidated the many different species and strains that make up this epidemiological complex, while other new methods have vastly increased our knowledge of the basic biology of the parasite. (Despommier's own contributions to our understanding of Trichinella have been pioneering and numerous.) For parasitology in general, it is safe to say that both the old and the new sciences will provide challenge and delight (and income) into the indefinite future.

Despommier's book, as its title tells, is about people. We can hope that readers, once hooked by these tales of human travail and attendant scientific exploration, will look further into the vast world of organisms that parasitize hosts other than human beings. Unless they do that, they are likely to forfeit the pleasure of knowing that there is a tiny flatworm, *Oculotrema hippopotami*, that lives in the eye of the host from which it gets its species name. That, of course, must seem a trivial matter (except, perhaps, to the parties involved), and indeed it counts for little in the context of parasitism and human affairs. But while there may have been a little truth in Pope's old aphorism that the proper study of mankind is man, there is a very large truth in the thought that the interest and concern of humans should extend to the creatures that share our planet, including those that we have domesticated—and of course their parasites. The price of a chicken dinner is determined largely by a single-cell parasite; the softness of our woolen sweater is affected by a whole suite of roundworms; and the durability of our leather shoes is diminished by parasitic larvae of insects. As long as people are omnivorous or carnivorous, the roundworms of cattle will (indirectly) be of greater importance to people than will the roundworms of people. And then there are the parasites of plants—including the parasites of the plants that nourish the cattle that nourish the people. Writer Carl Zimmer eloquently makes the point that parasites are in many ways drivers of organic evolution. Their relevance to our world may indeed be infinite! Despommier is well aware of these beyond-human realms of parasitology. Perhaps someday he will write a book about them. In the meantime we can relish his lively discourses on parasites, people, and places.

Writing from a medical perspective, Despommier properly talks about getting rid of parasites in people. Doing that might be OK. In the world at large, however, we must not think of doing such a thing. As Colin Trudge and others reminded us long ago, parasites (usually) fit neatly into the life of Earth. They are planetary treasures! As a properly selfish species (biologically speaking), we humans must protect our health, and the health of our domesticated partners. Ideally, we will not have to accept sickness; but we may have to accept the metabolic cost of parasitism as long as it remains subclinical. Since parasites are by definition harmful competitors in an interspecific union, we must find out how to control their numbers and their harmfulness. Balance, if we can achieve it, will keep us upright!

William C. Campbell
Drew University

PREFACE

"Successful systems attract parasites." So goes one of my favorite sayings about life. Our planet is literally covered with life of all kinds, from bacteria to elephants to the giant sequoias, and amazingly, each and every one of these organisms harbors another set of life forms: their parasites. Parasites are programmed to take advantage of other much larger life forms, feeding off their tissues. When we eventually go "where no man has gone before," I am convinced that we will discover life on other planets—and when we do, I have no doubt that parasites will be an integral part of their ecology.

By definition, parasites are needy. They literally can't live without us. As endearing as this may sound, cohabitation often comes with a heavy price tag: discomfort, illnesses, and sometimes even death. This may be why the concept of parasitism is a subject that at first has little appeal. Most of us are revolted by the mere thought that one of these creatures might someday find its way inside us. Imagine how it would feel to be told by your doctor that a worm the size of your favorite high-end fountain pen has managed to infect your small intestinal tract! It's the kind of thing that elicits nightmares. However, once you start learning the details of what these parasites are and how they carry out their sinister lives, it's difficult not to want to know more. These disease-producing organisms remain some of nature's most fascinating creatures.

One of the aims of my book is to reveal more of their lives to a wider audience than I have been accustomed to addressing. For many years I taught at a medical school, and students, academics, and practicing physicians were my only audiences. Now I want to share my thoughts with a wider readership.

Parasites come in a complete spectrum of sizes and shapes. I restrict my stories to those that possess a nucleus inside each cell. This group includes protozoa (single-celled organisms) and worms (helminths) but excludes the viruses, rickettsiae, and bacteria. I eliminate pathogenic fungi from the discussion, even though they too are eukaryotes.

The 1800s saw the establishment of parasitology as a bona fide biological science, and it arose in several places. In Germany this new field was spearheaded by G. F. Rudolf Leukhart, Theodor Bilharz, Carl Theodor Ernst von Seibold, and Arthur Looss. In England it was led by the likes of Thomas S. Cobbold, Patrick Manson, William Leishman, and Ronald Ross. By the early 1900s it had matured into a global science, bolstered by numerous new discoveries, including validation of the germ theory of infectious diseases and the role of arthropods (insects and such) in the spread of malaria. Today the field of parasitology continues to attract talented, dedicated researchers. Nearly every country has its cadre of experts who make their living studying them. As the result, a wealth of scientific literature now exists chronicling in exquisite detail the lives of a wide variety of these highly specialized organisms. Those interested in learning more can find many good books listed at the end of this volume. Several successful books written for the general public—including *New Guinea Tapeworms and Jewish Grandmothers* by Robert Desowitz and *Parasite Rex* by Carl Zimmer—and documentaries made for television, such as *Eating Us Alive* and *Monsters Inside Me*, have helped enormously to popularize the subject. I was privileged to be involved with the production of both of those Discovery Channel shows, and in a minor way with Zimmer's well-received book. I have also coauthored a book on the subject, *Parasitic Diseases*, now in its fifth edition, which is used as a reference for the medical profession but does not do the field justice when it comes to vividly portraying really juicy parasite stories. Hence the need for another sort of book in which the essence of the subject can shine without all the clinical jargon to water down the excitement of what parasites are and how they carry out their lives inside us. I hope to show by examples why we need to continue to find out more about what makes a parasite parasitic. If we do, some of the survival tactics of parasites may be able to help us, too.

Hollywood has picked up on the parasite theme from time to time. *Alien* is one of my all-time favorite examples of this. Everyone who has seen it remembers the showstopper: it comes early on in the film, when the "embryo" of the beast, in a torrent of blood and guts, bursts right out of the chest of John Hurt, then slithers off and raises havoc with the crew of that hapless spaceship. No matter how many times I've seen it, that "birthing" scene always freaks me out. But that was science fiction, and with some obvious inconsistencies in the parasitic life cycle, I might add. Trust me, the lives of real parasites are far more riveting.

Ever since we became a species, some 200,000 years ago, parasites have been responsible for untold amounts of human suffering and countless deaths. For instance, some experts believe that *Homo sapiens* almost became extinct as the result of epidemics caused by malaria that coincided with a time when our numbers were perhaps as low as 400,000 individuals. The worst part is that this killer is still with us. In just over the past one hundred years, as many people have died from malaria worldwide as now live in the United States. While the number of people dying from this one parasite is high, consider the fact that malaria in all its forms (there are four) infects some two billion individuals each year. This reduces the mortality rate to around 1 percent, making this group of infectious agents some of the most successful parasites on the planet. There are other evolutionary winners out there, too. For example, the number of humans currently infected with intestinal helminths (worms) is also in the billions. Tragically, what this really means is that there are a lot of people harboring more than one parasite.

It's not only humans; parasitism is also a very common theme among the rest of the world's flora and fauna. Most parasites specialize in a few select host species. Take beetles, for example, of which there are an estimated 350,000 known kinds (see Arthur Evans and Charles Bellamy's 1996 *An Inordinate Fondness for Beetles*). There are probably as many different species of parasitic worms that infect them.

Incredible as it seems, even bacteria are susceptible to the smallest versions of parasites, the viruses. These submicroscopic particles, referred to as phages, enter bacteria, replicate, and eventually kill them.

From the cool, measured perspective of a population ecologist, parasitism is simply nature's way of controlling the number of organisms that live in any given geographic region. Some biologists even go so far as to speculate that half the species on Earth are parasitic. But most of us are not ecologists, and our take on these freeloaders is to avoid them at all costs. For instance, we support through our system of taxes a robust sanitation industry, some form of which is found in virtually every community in the developed world. In most cases, that is all that stands between us and our parasites. If municipal facilities controlling the processing of wastewater and the safety of drinking water break down, we stand a good chance of suffering the ravages of a wide variety of infections and are forced to apply an armamentarium's worth of drugs just to keep them at bay until those facilities go back online. Unfortunately, the more than four billion people living in the less developed world cannot afford a functional public health system, or even the cheapest antiparasitic drugs, and are continually subjected to outbreaks of infectious diseases, many of which are caused by protozoan and worm parasites.

So despite all our efforts, the parasites still have the upper hand. The question is: Do we have to sit there and take it like all the other hapless host species on our planet? Or can we use our ingenuity and creativity to find ways of taking advantage of their molecular survival tactics? That is one of the main themes of this book. I am aware of many pieces of half-finished research on parasites that, if developed further, could benefit our species in some totally unexpected ways: breakthrough treatments of those suffering from type 1 diabetes, for example, or the successful transplantation of organs obtained from nonhuman animal sources, such as pigs. I also highlight the amazing ways in which these body snatchers succeed in carrying out their complex lives at our expense.

Much of the information I have drawn on reflects the new biology of the twenty-first century. Ever since the advent of the first crude microscopes, popularized by Antonie van Leeuwenhoek in Holland and Robert Hooke in England, we have become obsessed with knowing in every detail how parasites and all the other life forms on our planet carry out their lives. Over the past one hundred years of research in the life sciences, the complexities of biological networks and systems have become accepted paradigms for defining the ways life sustains and thrives in an ever-changing, often hostile environment. The closer we are able look at any living organism, the more we are drawn to understanding its complexities. We want to know it all. Today, science has provided researchers with a plethora of tools that they have then used to probe the inner workings of the cell, often without disturbing it. Florescent-labeled compounds that target just one aspect of its intracellular environment, coupled with sophisticated, high-power, computer-assisted microscopes give unprecedented resolution of cellular structures and their functions. The ease of sequencing entire genomes in just days has opened the door to a new world of investigation at the molecular level and will continue to offer exciting opportunities for a global understanding of the process of parasitism. Genetically manipulating organisms—inserting, altering, or even removing specific genes—in which the expression of individual molecules can be studied at will has helped to redefine what is and what is not essential for things such as infectivity, host range, metabolism, suppression of the immune responses, and site selection once they enter an animal. Many of the stories chronicled in this book could not have been written even five years ago, but owing to the rapid advance of cutting-edge technologies, parasitologists around the world are publishing heaps of new information that has altered forever our view of the host-parasite relationship. Hopefully, they will receive adequate levels of financial support to continue their work. All that having been said, at the molecular level, we are just beginning to appreciate how much we still need to know before we can say we really understand any given biological entity, be it host or parasite.

I hope that by the time readers come to the end of my book, they will have a new respect for an often misunderstood form of life and will also realize that if we ever succeed in eliminating parasites from our environment, we will still need to study their lives in the laboratory in order to be able to apply their strategies for survival to our own world.

ACKNOWLEDGMENTS

I could not have written this book without first having matured as a parasitologist, a process that continues to this day. I owe an enormous debt of gratitude to all my teachers and mentors who guided me along during my trek through this amazing field of biology. I might not have even become a scientist had I not been lucky enough to have had Dominic Casuli as my high school biology teacher, a gifted, caring, nurturing instructor. He saw promise in me and encouraged me to follow my passion for all things natural.

I am eternally grateful to Dr. Harold W. Brown, a brilliant physician/researcher who, at the very last moment of my frustrating search for employment before I would perhaps become actively engaged in the Vietnam conflict via the draft, found enough additional funds to hire me fresh out of college to work in his diagnostic/research laboratory. Dr. Brown not only probably saved my life, he deeply inspired me by his own commitment and passion for parasitology and the diseases parasites cause. He inspired me to continue my education at the graduate level.

Two researchers who greatly influenced me during my doctoral program at the University of Notre Dame deserve special mention: Bernard Wostmann, my mentor, and Morris Pollard, an amazing virologist who invited a budding parasitologist into his Saturday morning roundtable discussions to listen in on his graduate students' struggles with their own research projects. The insight gained at these meetings formed the basis for how I began to approach my own research problems.

I warmly thank William Trager, Miklós Muller, and Christian de Duve, three highly gifted and accomplished researchers at Rockefeller University with whom I had the extreme pleasure of interacting during my postdoctoral years. They generously allowed me to brainstorm with them, often on a daily basis. I also thank George Jackson for creating the opportunity for me to attend that renowned institution.

I warmly thank William C. Campbell, who has been an extraordinary figure in my professional and personal life. Bill continues to share with me the full breadth of his intellect, his resources, and above all his generosity and friendship. We have collaborated on research projects throughout our professional lives and have had an enormous amount of fun doing so. He read every word of my many drafts of this book. His numerous corrections and suggestions made the text read as it should. I take full responsibility for any errors that might remain undetected by either of us, and I give full credit to Bill for intercepting those that never made it into print. For that alone, I am grateful. I also thank Bill for agreeing to write the foreword.

I particularly thank Vincent Racaniello, a colleague and good friend at Columbia University, who single-handedly initiated the series *This Week In (Virology, Parasitism, and Microbiology)*. These podcasts may be found at virology. ws. He took time from his busy day to help me during the writing of this book, ensuring that my words and ideas would come together as they should in logical sequence, and with the correct take on the plowshares concepts. To this end, I also acknowledge the input of Steve Goff and Saul Silverstein, another couple of good listeners and colleagues who helped guide me to literature and ideas that improved greatly my understanding of the science of applied research.

Thanks go to John Boothroyd for reading the text carefully, then gently correcting any of my misconceptions regarding the biology of *Toxoplasma gondii*.

A most heartfelt thanks go to Patrick Fitzgerald and Bridget Flannery-McCoy at Columbia University Press for their help and encouragement throughout the writing and production phases.

There are many others whom I wish I could mention, but the economy of space on the written page does not allow for such luxuries. Nonetheless, I thank all of you with whom I have had passionate discussions and even arguments (too numerous to count) at various professional meetings, departmental retreats, Gordon Research Conferences, and so on and who will pick up this book and look for your names without finding them. Please know that your collective wisdom contributed greatly to my global understanding of the concept of parasitism and was invaluable in my rite of passage as a parasitologist.

Finally, I thank my wife, Marlene Bloom, for her patient and expert editorial input.

People, Parasites, and Plowshares

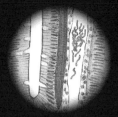

1 THIS NEW HOUSE

Trichinella spiralis

Live Long and Prosper

All parasites fall into just two groups. They are either short-lived (days to weeks), or long-lived (months to years). The first group holds little interest for me, since they have invested so little evolutionary creativity scheming up ways to live inside us without killing or seriously hurting us in the process. They simply get in and do their job; reproduction at all costs. In the end, either we kill them or they kill us, and that's that. The malarias (Plasmodia) fall into this category, and so do Giardia and *Entamoeba histolytica*, although these three can occasionally linger for up to several months in our blood or intestinal tract before we eventually fight them off. All these parasites happen to be single-cell organisms (protozoans). Granted, exposure to any parasite in this group could result in a protracted infection, but in the main, most of us manage to get rid of them in a relatively short period of time and go about our lives as if we had never encountered them. In contrast, there are some parasites that routinely stick around for months to years in our bodies and have acquired a toolbox full of chemicals and metabolic strategies that they employ to allow them to live long and prosper. It is this second group of organisms that never fails to hold my attention.

I spent the better part of my professional life, twenty-seven years, studying one of these long-lived creatures, and that is how I came to know and respect all of them for what they really are: superstars of the parasite world. Included in this long-lived group are the trypano-somes—both African (sleeping sickness) and American (Chagas' dis-ease) varieties—most of the leishmanias, *Toxoplasma gondii*, filarial roundworms (causative agents of elephantiasis and river blindness), the blood flukes (schistosomes), intestinal and tissue-dwelling trematodes, and the one I chose to study, *Trichinella spiralis*, the nematode respon-sible for trichinellosis (trichinosis).

Trichinella spiralis is a nematode that infects mostly mammals, but other members of the trichinella family infect birds and reptiles, too. Let's take a moment to briefly describe what a nematode is before diving into my favorite subject. All nematodes are nonsegmented roundworms and include both parasitic and nonparasitic examples. In contrast, earth-worms are segmented and nonparasitic. In fact, most nematode species are free living and are found in soil. Their ecological role is to help estab-lish healthy growing conditions for most plant species. All nematodes, be they parasites or not, have plenty of biology in common, including their basic body plan—a nonsegmented tube consisting of a chemically resistant, tough outer coating called the cuticle; muscle cells for locomo-tion; a nervous system for sensing where they are in the environment; a robust reproductive system for producing lots of offspring; and an ex-cretory system for handling liquid metabolic wastes. The great majority of nematodes are either male or female, and they achieve adulthood by a four-stage developmental program in which each progressive stage is preceded by shedding their cuticle to make room for the added growth and development of the next stage. Adult nematodes do not shed their cuticles, as they have achieved maximum growth by then.

Trichinella has a complex life cycle that is all about survival. It does not have a reputation as a killer, although it is fully capable of doing so if the victim eats enough raw meat (often pork, but the parasite is found in other meats too) containing an abundance of the larval stage. All trichinella really wants to do is infect a host and then be carried far away from the site of infection. So keeping the host alive for as long as possible is in its best interest. When an animal harboring the infective stage of trichinella eventually dies of something else, perhaps old age or at the hands of a predator, mammalian scavengers that consume the carcass will become infected, thus extending the geographic range of the parasite. As a result, some form of trichinella (there are currently eight known species) can be found in almost every corner of our planet, including such disparate locales as the jungles of Papua New Guinea, the Serengeti Plain of East Africa, and the Arctic tundra.

Most trichinella species infect more than one kind of host—black bear, cougar, walrus, hyena, human. *Trichinella pseudospiralis* infects birds and mammals. Trichinella occasionally even infects herbivores, since most grazers will eat meat when the opportunity arises. Parasitologists call this kind of generalist infectious agent "promiscuous" because of its adaptation to a broad range of host species. Trichinella will even infect bats, but this only occurs under laboratory conditions. The only mammals in which trichinella has not been found in nature are unusual ones: moles, shrews, anteaters, pangolins, echidnas, and platypuses, for example. All trichinella species infect the gut tract of their hosts as adult worms and produce live larvae that find their way to the striated muscle tissue, where they can live for up to twenty-five years. Amazingly, trichinella larvae can even live weeks beyond the life of its host—quite a remarkable feat when you think about what it means to be a parasite. So how does this ubiquitous worm engineer its own survival once it is eaten? Before I reveal some of its biological secrets, I'll first describe how it was discovered, since it is this bit of history that kick-started the search for more parasites and put *Trichinella spiralis* on the map as a human pathogen.

Seeing Is Believing

The place was London, England, and the year was 1835. The last outbreak of the Black Death had occurred in 1665, and the streets of London were once again filled to the brim with the hustle and bustle of
commerce and people. Nonetheless, the horrendously unsanitary ur-
ban environment that favored the spread of plague—raw garbage in the
streets, with free-range rats and their fleas roaming around—had not
changed much since those dark days. The rich managed to stay on their
somewhat cleaner side of town, while the disadvantaged masses crowed
into the filth-ridden downtown districts near the River Thames. Charles
Dickens would later write vividly about all this in his classic *A Tale of Two
Cities* (1859). Cholera would not arrive in London by boat from India
for another two years, but other endemic diseases continued to spread
and raise havoc with the general population. Tuberculosis and syphilis
were rampant. Science-based medicine was in its infancy in 1835, and
the germ theory of infection (the one that proved that microbes are the
cause of many common illnesses), brought to light by gifted researchers like Louis Pasteur and Robert Koch, had yet to be demonstrated.
So the actual cause of yellow fever, malaria, cholera, plague, syphilis,
or tuberculosis remained a complete mystery. Nonetheless, with each
new medical discovery, establishing a career as a physician became an
increasingly attractive option to well-educated members of the public.
It attracted James Paget, and thus begins our story.

We know who James Paget was because he became the most famous
pathologist of his time and described in detail numerous chronic diseases, including the one that still bears his name. Later in his illustri-
ous career he was knighted by Queen Victoria. However, his connection
with the discovery of the worm we now call trichinella occurred when
he was just a first-year medical student at St. Bartholomew's Hospital,
located in West Smithfield in London. All the city folk referred to the

place simply as St. Bart's, and I will, too. It is still there today and remains one of the finest hospitals in the United Kingdom.

It was a rainy Monday morning that February 2, as young James, along with his fellow medical students, stood in a tight semicircle in the autopsy room at St. Bart's, observing the dissection of a 51-year-old bricklayer who had apparently succumbed to the ravages of consumption. Nothing unusual there, as death from tuberculosis was a common occurrence, not only in London, but throughout Europe too. The surgeons who were demonstrating that day were not pleased by what they encountered, but TB was not the cause of their discontent. They began the autopsy around 11:00 a.m., using beautifully handcrafted scalpels that were sharpened and maintained by the medical technicians back at the main hospital. As the pathologist/surgeons commenced to lay open the corpse, they immediately observed that the pressure they had to exert with their instruments to cut into the chest cavity was more than what was usually required. In fact, their instruments were being dulled by the simple act of drawing them across the unbroken skin and into the superficial muscles of the deceased construction worker. Swearing and cursing ensued, followed shortly thereafter by the throwing of dull-edged scalpels to the floor in disgust. Finally, after gaining access to the internal organs of the thoracic cavity and inserting his hand, one of instructors was heard to mutter, "Just as I thought. Another @#$%$#@# case of sandy diaphragm!" We know all this because young Paget kept an exquisitely detailed diary in which he wrote every day about the goings-on in medical school. He also wrote letters to his brother from time to time, especially when something unusual came his way. As it happened, this was to become one of those unforgettable moments.

The entourage of surgeons apparently had had enough, dulling nearly all their scalpels without much to show for it. "What causes sandy diaphragm?" Paget asked, but no one replied with the answer. Apparently, the surgeons at St. Bart's were not terribly curious when it came to things outside their own narrow sphere of interest. Finally, in a fit of frustration, they abandoned the lesson altogether and headed straight out the door and off to lunch. The students followed suit—with the exception of a professor, Thomas Wormald (a most propitious name, under the ensuing circumstances), who had dutifully remained behind to help clean up the mess and make room for the next corpse. After everyone but Wormald had left, Paget circled back and snuck in to remove a small sample of the diaphragmatic muscle tissue, seeking satisfaction of his earlier question. Paget was odd man out in that regard, and his insatiable appetite for knowledge drove him to look for answers himself. We are forever in his debt for doing so, because in the end he succeeded.

The first-year medical student was in the habit of carrying a hand lens with him, and he used it now to partially satisfy his curiosity. James observed what he thought looked like tiny, wormlike critters within the small, white, sandy structures scattered throughout the piece of dia-phragmatic tissue. To make sure of his observations, though, he needed to look at the sample using a microscope, so he hurried over to the British Museum of Natural History where one of those rare instruments was located. The one James knew about belonged to Robert Brown, a botanist, who himself became famous for identifying "Brownian motion," a characteristic of small particle random motion that would later contribute to Einstein's theory of relativity.

When Paget peered into Brown's instrument at the snippet of muscle tissue, his original suspicion was confirmed. Worms indeed were the cause of sandy diaphragm. On February 6 he gave a small presentation on his discovery to the student club at St. Bart's. In addition, he wrote his brother a letter, describing in some detail the worm and its surrounding tissue. So apparently Paget was the very first human ever to lay eyes on trichinella. Or was he?

When Paget had first removed the sample from the body and examined it with his hand lens, Wormald—who was still in the autopsy room—confronted him and demanded to know what he had just seen. The fledgling pathologist replied that he thought he had seen worms but needed to confirm it using a better instrument. Then and there, Wormald decided to take the issue of sandy diaphragm to a higher level. After James headed off to the museum, Wormald cut out another small piece of muscle tissue from the same anatomic location and rushed it over to his good friend Sir Richard Owen, none other than the director of the very same British Museum of Natural History. Owen had a better, more powerful microscope than Brown's and ultimately saw more details of the worms and tissue structures than Paget had observed. That day, Owen sat down and made a series of drawings of what he had seen and hurriedly wrote a paper on it, submitting it to the Royal Society for presentation. He knew of Paget's original observations as described to him by Wormald but gave the medical student little credit in his tome on the subject. Owen named the parasite *Trichina spiralis* because the hairlike worms were coiled up in a spiral configuration inside their microscopic homes, which Owen referred to as "cysts." The parasite's name was later changed to *Trichinella spiralis* when it was discovered that another organism had already been designated Trichina (Greek for "little hair").

The Rest Is History, Too

Some of you may recall that it was also Sir Richard Owen who vehemently opposed Charles Darwin's concepts of how species arise and used a religious argument as his main strategy of attacking Darwinian reasoning. Owen used every public opportunity to dissuade the listeners as to the validity of that world-changing idea. He even went so far as to surgically alter the brain of the only known lowland gorilla to make it look anatomically different from that of a human brain, thereby "proving" Darwin wrong once and for all as to our apelike origins. It was up to Thomas Huxley, known as "Darwin's bulldog," to get his hands on a second gorilla brain and show that it was indistinguishable in every way from that of a human's. Shortly thereafter, Owen was discredited in a public debate with Huxley. He then resigned his position at the museum and retreated from science altogether, never to be heard from again. To this day, however, Owen is still given the lion's share of the glory associated with the discovery of *Trichinella spiralis*. In 1996 I visited that fabled museum and inquired as to the whereabouts of the original sample of diaphragm from which Owen made his observations, but I was told by the curator of helminths that, alas, a German bomb had destroyed the Duveen Gallery in 1940 with the specimen in it.

In the ensuing fifty years following its discovery, many eminent scientists worked on elucidating the details of trichinella's life cycle, examining its medical relevance to disease, and exploring the epidemiology of how this unique nematode parasite spreads from animals to people. Today we know a great deal about all phases of its life, and with the completion of the sequencing of its genome in 2011, we will undoubtedly know much more in the coming few years. In the meantime, I will summarize its complex home-building activities in the host's muscle tissue, and show how we might take advantage of this knowledge to solve some of our own medical problems.

Life in the Raw

As mentioned earlier, trichinella is not just one species as originally thought, but rather eight separate ones, and we will probably discover new members of that worm family the more we look for them. However, I will only discuss the biology of *T. spiralis*, since it is the most commonly occurring member of the trichinella family, and therefore the best studied. It is also the one most responsible for infecting humans.

Its life in a new host begins with the ingestion of raw, infected meat (fig. 1.1). While it is in the meat, it is a larva or immature worm, the one Paget and Owen first saw. It is encapsulated inside a specialized host cell termed the *Nurse cell*. It is this structure that calcifies in old infections (thirty years in humans), causing the "sandiness" that the St. Bart's surgeons dulled their scalpels on. Their blades were, in fact, encountering the petrified homes of the parasite.

That we can become infected by eating meat bought at the grocery store says a lot about how this worm is able to survive in the wild as well. After all, ground pork does not qualify as a living entity! In nature the worm is able to hunker down and ride out the time it takes an average carcass to become consumed by mammalian scavengers, such as bears, coyotes, or wolves. Larvae can survive for up to a month in a dead animal at low temperatures. Animals that hibernate have a natural antifreeze molecule to help them survive the cold winters, and this same molecule also protects some species of trichinella, if the worms should also become frozen. *Trichinella spiralis* is an exception, luckily for us, and is easily killed by freezing.

After an infected piece of meat is swallowed, the worms are released from their Nurse cells by our stomach acid and by pepsin, a digestive enzyme that dissolves the meat portion of the meal but spares the parasites. Trichinella is able to survive because it is covered with a resistant coating, which allows it to pass unharmed from the stomach into the small intestine. There the immature worm penetrates into a row of intestinal cells and begins its journey to adulthood by shedding its outer cuticle. It does this three more times and emerges as a much larger adult worm some thirty-six hours later. At this point the worms have matured into males or females. Mating ensues in the lumen (the hollow space) of the small intestine. The inseminated females then reenter the intestinal cells and five days later begin to give birth to live offspring. The males, having fulfilled their part in the life cycle, do not reenter the intestinal cells and are carried out of the host with the fecal matter.

Trichinella spiralis

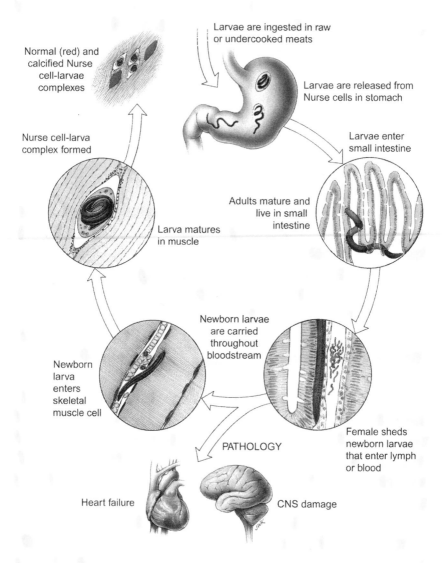

Figure 1.1 *Trichinella spiralis* life cycle. Illustration by John Karapelou. (*Source: Parasitic Diseases*, 5th ed., (c) Apple Tree Productions, LLC, New York)

House Hunting

The active newborn larvae (fig. 1.2), penetrate the gut wall and enter the bloodstream to begin their own journey throughout our body. The heart pumps infected blood to all the organs, and the newborns eventually get stuck in capillaries because their diameter is wider than those vessels. This triggers the immature parasites to move out of the bloodstream and, using their oral spears, to burrow into the surrounding tissue. If they enter striated skeletal muscle (most mammals are nearly 40 percent muscle), they are well on their way to completing their portion of the life cycle (fig. 1.3). If they penetrate organs—heart, kidney, liver, lung, brain—they simply penetrate back into the circulating blood and continue their search for muscle bundles.

The worm has the ability to detect just where it is in the host and act accordingly. That is because it has a small, well-organized brain connected to a peripheral nervous system that it uses to make sense of its immediate environment. This searching behavior damages the tissues

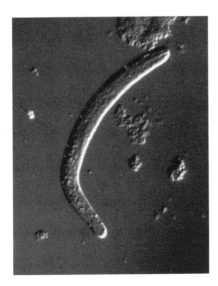

Figure 1.2 Newborn larva of Trichinella spiralis

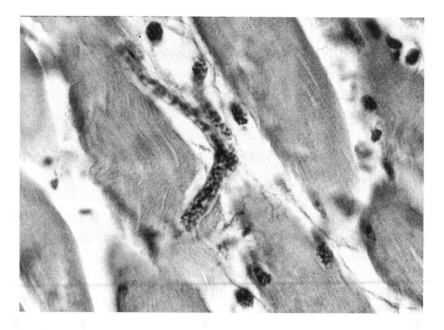

Figure 1.3 Newborn larva entering a muscle cell

and in humans often leads to clinical disease. If enough newborns are produced (each female worm can give birth to about one thousand offspring), the infection can be deadly. All this depends on how many immature worms are eaten and how large the host is. Trichinella infection in adult humans has a 5 percent mortality rate.

The Frank Lloyd Wright of Parasites

Once it encounters a skeletal muscle bundle, the worm penetrates into one of that tissue's single cells and initiates a remarkable process: the formation of the Nurse cell (fig. 1.4). I have always thought of trichinella as a master builder, along the lines of a Christopher Wren or Frank

Lloyd Wright. The analogy is worth spending a moment on. First of all, these iconic, talented architects never actually laid their hands on a project. Rather, each one imparted his creative skills to the design and execution of a series of structures, all of which speak to their ability to see an empty space and fill it with beauty and functionality. In that sense, this unusual worm behaves similarly, taking a muscle cell (empty space from the worm's perspective), and turning it into a fantastic, co-coonlike living structure, whose sole purpose is to allow the parasite to grow up into the infective stage and to remain in the body as long as the host remains alive.

The Nurse cell is truly a thing of beauty, and a biological entity far more functional than any skyscraper. In 1976 a close friend of mine, Eric Gravé, a gentle man and a most talented microscopist, took a

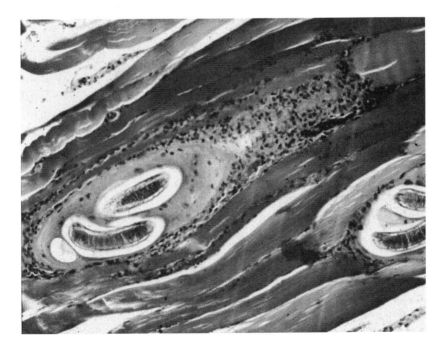

Figure 1.4 Nurse cell-larva complex

Figure 1.5 Nurse cell-larva complex. (Courtesy Eric Grave)

photomicrograph of a single Nurse cell I had isolated and preserved for him. He entered it in the Nikon Small World Contest and won first prize (fig. 1.5). Granted, knowing that a parasite, a worm no less, has commandeered one of our cells to aid in its survival is an abhorrent idea to most of us, but the judges of that prestigious competition set aside their personal prejudices and voted unanimously to recognize the beauty in nature, regardless of its intent.

Other similarities between trichinella and the construction industry have also occurred to me from time to time. I will never forget the circumstances surrounding a particular lecture I gave at Woods Hole at the Biology of Parasites course. I entitled my talk *"Trichinella spiralis: This Old House Revisited."* I had no idea that most of the students in attendance that day had come from Europe and the UK. I had to spend several minutes describing the premise of that popular television show before they caught on to what I was trying to convey regarding the tasks that the worm tackles: roof building, gutting the interior of an old "house" and fabricating a new one on the same architectural footprint by reusing materials derived from the old one, and so on. They all then smiled and began to understand my approach of introducing them to the concept of Nurse cell formation.

This New House

In fact, home building is exactly what the newborn larva sets about doing immediately after it enters its new intracellular domain. The entire process takes about twenty days to complete. It should be mentioned here that only five of the eight species of trichinella induce the host to assemble a completely formed Nurse cell. The others engineer partial versions of it. What follows applies only to the first group. The very first thing the worm does after penetrating the cell is to crawl away from the point of entry. The cytoplasm then collapses around the damaged portion of muscle cell and seals off the hole, preventing precious cytoplasm and intracellular fluids from escaping. At this point the parasite is now completely surrounded by a latticework of scaffolding (e.g., myosin light and heavy chains, sacroplasmic reticulum), all of which is committed to aiding the host cell in doing work. The parasite has no use for these existing structures, so it sets about directing the host to make radical changes in its new milieu by secreting biologically active substances into its immediate environment through its mouth. These parasite products cause the structural elements in the immediate vicinity to dissolve. Since the muscle cell is quite long compared with the parasite, most of the contractile elements are spared, allowing the host to continue to function almost within normal limits. The mechanisms the parasite employs next to complete the job of home building involve altering the normal patterns of gene expression of the host cell, but molecular parasitologists are not yet sure of the exact ways in which it does it. Understanding these steps in the process will be important in helping to realize not only how this unique parasite survives for long periods of time inside us but also the ways our own cells respond to a variety of chemical influences foisted on us by our pathogens.

Within several days, after the structures associated with muscle function have been destroyed, new host-derived structures arise to replace them. All this activity generates a pool of metabolic resources—amino acids and other small molecular building blocks—that provide food for the worm. Around five to seven days after entering the muscle cell, the parasite begins to grow and mature. In doing so, it remodels its own cells and makes new proteins that it first stores in several types of granules in specialized cells called stichocytes (Greek for a row of cells). It changes the contents and shape of its now empty row of newborn larva-specific secretory cells into a new row of cells, each of which contains a unique array of granules that differ from those in the newborn larva. The stichocyte granules contain proteins that represent the next set of blueprints for constructing the parasite's new house on the old foundation. These new granule types come in five distinct varieties, easily identified under the election microscope by their shapes and sizes. Every granule type contains a different set of proteins, each of which is the actual message from the architect (the parasite) to the workers (the host genes). At various times after they are made, these parasite instructional molecules are exported into the matrix of the early Nurse cell, and each plays a central, specific role in the generation of a multitude of structures associated with the mature parasite–Nurse cell unit. Around day eight after it enters the muscle cell, it begins to export these newly made proteins into the surrounding host tissue. This forces the muscle cell to change into a Nurse cell. Again, I emphasize that none of these new structures are made by the parasite, which only directs the action.

Mirror Image

Kaufmann House in rural Pennsylvania, known to most as Fallingwater, was designed and executed by Frank Lloyd Wright. It is a marvel of modern architecture and thus worthy of comparison to a Nurse cell in muscle tissue. The house is integrated into the middle of a dense forest, and a trout stream, Bear Creek, actually flows through the center of its footprint. It is so tied into its natural surroundings that a flash flood in 1956 nearly washed it away. The stream doesn't just soothe the inhabitants with its babbling sound; it actually provides them with water. Wright was obsessed with these kinds of integration issues. While he was not always successful, he never veered even one degree from his goal of bringing people back in synch with their natural environment. He reasoned, correctly, that establishing a natural connectivity between the built environment and the surrounding landscape would reward those living there with a more enriched and serene life.

Similarly, the Nurse cell's raison d'être is to insure a safe and sustainable life for its occupant, which is no mean feat considering that the host would much prefer to be parasite free. It is called the Nurse cell because of this singular trait; it "nurses" the worm from a newborn larva to its infective stage, then maintains itself and the parasite until another host ingests it. This specialized cell is located in a wide swath of normal host muscle tissue, the forest equivalent to Fallingwater. While the Nurse cell-parasite complex has no floods to contend with, it does have other dangers to avoid. For example, it has to keep host immune responses at bay, which have the potential to end the life of the worm. So immunosuppression is a major clause in the contract between the host and trichinella. It also has to ensure that the worm gets properly fed and is able to get rid of its wastes in a safe and efficient way. To this end, the parasite instructs the Nurse cell to recruit a blood supply, starting about eleven days after the worm enters the muscle cell. This could be seen as the stream in the forest, if you will. By that time, the parasite inside has increased in size some fivefold compared with the newborn larva and has recycled all the host-derived molecules from its demolition phase. For the rest of its life, trichinella is now totally reliant on nutrients it obtains from the host's bloodstream.

I Love You Just The Way You Are. Now Change.

Inside the Nurse cell, the environment is remarkably different from that of a normal host muscle cell. For one thing, the parasite is now strictly anaerobic. It consumes no oxygen. This means that the internal environment of the Nurse cell must support this type of metabolism. Thus blood vessels recruited to the Nurse cell cannot be arterioles or arteries because these are the ones responsible for bringing oxygen-laden red blood cells to the tissues. Rather, the vessels that form the network around the Nurse cell are sinusoids. These are the same ones that facilitate the entry into the blood of all endocrine gland-derived hormones (e.g., insulin, adrenalin, vasopressin, thyroxin). Because these specialized vessels are leaky compared with capillaries, they allow large molecules such as insulin, produced in the pancreas, and albumin, synthesized in the liver, to freely enter the venous circulation and return to the heart for rapid distribution to the entire body via the arteries. Thus learning more about the molecular details of how trichinella engineers its host to generate a Nurse cell might have real value as a strategy for treating a commonly occurring serious disease—details on that in a moment.

It's Not Just About Me

If the infection lasts long enough, some of the Nurse cell–parasite complexes die and ultimately calcify. Our 51-year-old bricklayer had probably contracted his infection at a young age, say between 14 and 20. This would have given his Nurse cells plenty of time to mature and carry out their job of maintaining the parasite. Eventually they began to die, and the result was "sandy" diaphragm. So what is in it for us? Do parasites simply take everything without conferring any benefit on the unfortunate host? This may sound surprising, but perhaps some of them might allow us to enjoy a slightly longer lifespan—because by giving the host an extra boost to live longer, the host will be able to venture forth to spread the worm to new places.

The following anecdote could be scripted right out of a Hollywood melodrama, but it is true nonetheless. You may be familiar with the renowned actor Yul Brynner. He starred on Broadway in the The King and I. In film Brynner was equally well-known, with featured roles in The Ten Commandments, Taras Bulba, The Magnificent Seven, and Westworld, to name a few of his more popular ones. In 1973 Brynner and three dining companions all contracted trichinellosis at a popular restaurant, Trader Vic's, located in the posh Plaza Hotel in New York. A year later he won a settlement of $125,000 in damages. If he knew that the infection might have been responsible for prolonging his life, even a little, he might have thought twice about the litigation. Tragically, Yul was also a heavy smoker and was diagnosed with lung cancer. He died several years later from it on October 10, 1985. I think that it is quite possible that his trichinella infection slowed down the growth of the cancer, since this worm is known to activate all aspects of our immune system, except those responses that might act directly against the Nurse cell. Data from laboratory animal experiments using a wide variety of other maladies plus infection with trichinella strongly support this view. Tumors of certain kinds regress, and many infectious agents that would ordinarily make a host sick do not cause disease in those infected with trichinella. BCG vaccine (a killed preparation of the bacillus of Calumet and Guerin, a TB-like microbe) has similar effects on our immune system, placing all aspects of it on a higher level of alert surveillance for things like cancer cells and pathogenic micro-organisms.

Blood Is Thicker Than Water

Trichinella offers us other clues that could be of value. For example, the parasite may provide a new approach to the treatment of type 1 diabetes. In this insidious illness, the patient's pancreatic islet cells fail to continue the production of insulin. The reasons for this are poorly understood, but one viable hypothesis suggests that sometime in our childhood, we encounter a viral infection whose antigenic signature closely resembles that of the islet cells of our pancreas. When we fight off the virus, we are left with an immune system that now turns its attention to our own tissues, since the infectious agent is no longer present. This kind of disease is referred to as autoimmune. Standard treatment for juvenile diabetes involves frequent injections of insulin. Failure to do so has dire consequences. But, unfortunately, the treatment is not perfect, even when the patient is diligent. Eventually small excesses of sugar in the blood stimulate the formation of unwanted new capillaries. When the heart is unable to efficiently pump blood to all parts of the body because of the extra vasculature, the peripheral circulation begins to fail. The loss of fingers and toes may occur, followed by renal failure and loss of sight. Eventually death ensues. This dire prognosis means that other treatment approaches are desperately needed.

Transplanting pig islet cells into diabetic patients is one treatment that has been pursued, and it seems like a viable alternative if certain issues could be addressed. Early human clinical trials with this treatment were promising but quickly encountered several hurdles. The first was, of course, the fact that a pig donated the cells. Most of the time our immune system is really good at distinguishing between us and everything else, so immunosuppression of the recipient is necessary. The second hurdle proved far more difficult to jump over. When individual islet cells were injected intravenously, they became stuck in the patient's capillaries, just like the newborn larvae of *T. spiralis* did during their migration in the bloodstream. Oxygen levels dropped in the affected area, triggering the synthesis of a new section of capillary that bypassed the blockage of the circulation at that point, restoring blood flow. This left the islet cells stuck on the tissue side of the circulatory system surrounded by nonleaky vessels, so when blood glucose levels went up after the patient ingested a meal, the islet cells were slow to respond. Without a mechanism for rapidly getting insulin into the bloodstream, the endothelial cells that form the capillaries in all the tissues ate up the excess sugar and multiplied, creating a tangle of new vessels. In the end, islet cell transplants slowed down the overall pathology of type 1 diabetes, but the disease eventually won out and the patients returned to an acute state of illness.

Plowshare Concepts

How can a worm that lives a long time in a portion of an altered muscle cell help these diabetic patients? Recall that trichinella induces sinusoids during Nurse cell formation, and that these vessels are associated with the endocrine gland system because they allow hormones easy access to the venous circulation. What if we were to learn that trichinella induces its sinusoids by secreting a specific protein that, in the infected host, initiates the metabolic program that generates that particular type of vessel? Using a molecular biological strategy, we could then search the worm's genome for the gene encoding our desired protein and clone it into pig embryos. By connecting a molecular on-off switch, known as an inducible promoter, to that gene so we can control its expression, we could then let these special pigs grow up. Whenever we desired, we could isolate their pig pancreatic islet cells and inject those engineered cells into our patients. Instead of eliciting capillaries, we could turn on the switch for the trichinella gene responsible for making sinusoidal vessels. Now the insulin hormone molecule can easily traverse the vessel wall and rapidly enter the venous return. We would still need to mildly immunosuppress the patient, but now they could function on their own without the need for injecting insulin.

We may be a long way from this achievement, but because we have determined the entire DNA sequence of trichinella's genome, anything seems possible. The first step would be to identify and then isolate the protein responsible for the induction of sinusoids. We have also determined the DNA sequence of our own genome. Proteomics, the new field of biochemistry established as a partner with genomics, is dedicated to describing all the proteins that an organism's genes encode, the key to achieving this goal. For example, we have learned that our own genome encodes around twenty-seven thousand different proteins. We need to describe each one of those proteins before we can begin to make sense of how our genes work to produce an organism as complex as a human being. Since all other life on Earth has a similar genome, we aspire to describe all their genes and all their proteins, too.

Inserting our gene of choice into the pig embryo has been worked out for many other systems and does not present any technical barriers that need to be overcome before proceeding. Completing the rest of the story would involve many clinical trials and lots of hard work on the part of clinicians. Fortunately, we scientists are up for the task. All we need to do is obtain funding for the project; no easy feat in today's world of tight National Institutes of Health budgets and targeted research programs, and therein is the rub. High-risk proposals such as the one I have suggested fall well outside the guidelines of traditional funding streams, but the rewards would be great if they were to succeed. Here's a good example of where our reach should exceed our grasp. If we learn to listen closely, the parasites will share their innermost secretes with us.

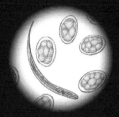

2 HOOKED ON PARASITES

Ancylostoma duodenale and *Necator americanus*

The War Between the States had been over for some forty years, but reconstruction was going much too slowly, especially for John D. Rockefeller, Sr. Each year he would ride in his sumptuously appointed private rail car, sometimes with a few close friends (Thomas Edison and Henry Ford, for example), to Florida for his annual winter vacation. Rumor has it that all the way down and back, the group would commiserate about the southern white folk's apparent lack of ambition. These captains of industry adhered to the belief that it was this singular behavior that was responsible for holding back the southern economy; a widely held hypothesis firmly grounded in Union prejudice against anything Confederate and fueled by a Protestant ethic that literally demanded of its practitioners utter distain for the nonindustrious portion of humanity.

There was one piece of evidence in support of their view, and it was based on their personal observations. En route to their winter compound just outside Tampa, once they had traveled well below the Mason-Dixon Line, they saw more than a few people idly sitting on their porches in their rocking chairs, apparently with nothing better to do than to watch the trains go by. Why weren't these people out working and contributing to the economy? For Rockefeller and his companions, the answer was clear; they were just plain lazy. Their now abandoned cotton plantations had been made possible solely by the manpower supplied by slaves. The southern landed gentry did little if any physical labor before the Civil War, so obviously no matter what the situation, southern whites had no stomach for an honest day's hard work. But these Yankee tycoons' simplistic rationalization as to why the South had not risen again was dead wrong, as history would soon show.

Back in New York, Rockefeller decided to get to the bottom of the problem once and for all. The reason was simple: greed. He needed to improve his own economy, and taking advantage of new markets was his modus operandi. The South was ripe for exploitation, if it could only gain some economic momentum.

Meanwhile, the science of parasitology was emerging from the nineteenth century as a full-blown field of study. America had its early heroes, including Charles Wardell Stiles, who took hookworms (tiny, gut-dwelling, blood-sucking worms) for his research subject. He was following the lead of his mentor in Germany, Arthur Looss, the acknowledged world expert on that parasite. Looss was a gifted experimentalist as well as an accomplished epidemiologist. But it was a fortunate accident rather than scientific trials that would reveal the secret of how this parasite got transmitted from one person to another.

While working in his laboratory, Looss accidentally came in direct contact with a culture of hookworm larvae, a 50/50 mixture of feces containing the eggs of the parasite and garden soil. Within several days of the culture being set up and maintained at room temperature, the eggs developed into infectious larvae. Each day thereafter, as they would in nature, hundreds of immature worms migrated to the surface of the culture dish and gathered into large aggregates. En masse, the bundles of larvae then began slowly and rhythmically to wave back and forth in unison, a movement that resembled a miniature flickering flame. This behavior is known as "questing"; the parasites are actively looking for a host. Up to that point, no one knew anything about how the infection was contracted. Looss noted that several hours after accidentally touching the culture's surface with his hand, the point of contact began to itch. He suspected that he had become infected. At that moment, he may well have reflected on that now famous saying by a contemporary infectious disease legend, Louis Pasteur: "Luck favors the prepared mind." So to find out for sure, he began to inspect his own stool for the presence of hookworm eggs. Lo and behold, within several weeks, eggs appeared in his feces. This revealed the portion of the life cycle that involved the penetration of unbroken skin by the infectious larvae, and the time it took to produce fecund adult parasites (also known as the period of patency). Physical contact was all that was necessary to become infected, unlike so many other infectious disease agents that had to be either swallowed (e.g., the dysenteries, cholera), transferred by sexual activity (e.g., gonorrhea and syphilis), or transmitted by a blood-sucking arthropod vector such as a mosquito (malaria), tick (Lyme disease), or blackfly (onchocerciasis).

After studying with Looss in Germany, Stiles returned to the United States and began to look for hookworms in America. He was not disappointed. Stiles showed that hookworm infection was in fact endemic in many parts of Texas and the southeastern United States. But this worm differed slightly in its morphology from the one Looss had inadvertently provided a home for. Stiles named it *Uncinaria americanus*, the American hookworm. It should be pointed out that Stiles was not correct about the global distribution of *U. americanus*, as later studies would show. It is also found in Africa and was presumably introduced into the southern United States via the slave trade. This fact was to prove crucial to the development of a new theory as to why the Confederacy lost the war, but more about that later on. Stiles suspected that it might be responsible for causing clinical illnesses, but he had no direct proof to back up his suspicion. As luck would have it (bad for Italian tunnel workers, good for Stiles), an outbreak of anemia in 1880 among Italian laborers engaged in digging the Saint Gotthard Tunnel was to provide the key. Edorado Perroncito, a veterinary pathologist in Turin, had shown that infection with a related species, *Ancylostoma duodenale*, induced anemia and, with heavy infection, could even result in death. He based his conclusions on the coincidence of the anemia epidemic in the tunnel workers with his autopsy findings on several of the fatal cases, in which he found abundant numbers of adult *A. duodenale*.

This outbreak gave Stiles the indirect evidence he was seeking to make a case that the American hookworm was causing clinical illness in the American southland. Stiles's results reinforced his view that "southern laziness" might actually be caused by an inconspicuous little worm with a growing nasty reputation. In 1909 Stiles laid out the details of his hypothesis to Fredrick T. Gates, a longtime trusted adviser to Rockefeller, who then related it to Rockefeller. The theory had enough circumstantial evidence and believability to convince Rockefeller to convene the Rockefeller Sanitary Commission for Eradication of Hookworm Disease (RSCEHD). It consisted of eleven highly selected individuals with a wide breadth of experience in management, education, and the biological and medical sciences: William Henry Welch, Simon Flexner, Fredrick T. Gates, Edger Gardner Murphy, Edwin A. Alderman, H. B. Frissell, Walther Hines Page, John D. Rockefeller, Jr., James Yadkin Joyner, Philander Priestly Claxton, and David F. Houston. The senior Rockefeller donated one million dollars out of his own fortune to the effort, and off the commission went.

Once its members began their investigations into the pathological basis of southern anemia, it became obvious that it was not caused by malaria (the most common cause of anemia known up to then), due to that mosquito-borne parasite's characteristic acute onset of symptoms and seasonality of occurrence. At the urging of Stiles, microscopic examination was conducted of stool samples from people obviously suffering from chronic anemia, which quickly confirmed that they were all heavily infected with hookworms. The parasite that Stiles had identified in the United States was then renamed *Necator americanus*—the

American killer. It seems that parasitologists had developed a flair for the dramatic.

If you want to get an idea of what an adult who has had severe hookworm disease as a child looks like, rent the movie *Deliverance*. The banjo player, portrayed by actor Billy Redden, is a dead ringer for those most affected by the "germ of laziness." Their growth and development have been compromised, and they have permanent intellectual impairment. I would like to think that Redden has none of these other characteristics! Discovering who these "wormy" people were and treating them on the spot reduced the so-called community worm burden, ensuring that future generations of children living in the same area would not incur heavy worm burdens, and thus sparing them from developing the disease. Even Mark Twain commented on the findings of the commission: "The disease induced by hookworm . . . was never suspected to be a disease at all. The people who had it were merely supposed to be Lazy, and therefore were despised and made fun of, when they should have been pitied" (*Letters from the Earth*).

So "southern laziness" translated into a knowable disease whose main feature was iron-deficiency anemia. Thus began a major overhaul of the sanitary infrastructure in America. With the help of the commission, each of the most affected southern states instituted a multilevel control program specifically established for the eradication of the *American killer*. The program consisted of public health education, installation of latrines (outhouses situated over a 6-foot-deep hole), and a mandate that all school-age children wear shoes of some kind. The outhouse regulation regarding the depth of the hole was based on simple experiments that had shown the crawling range of the infectious hookworm larva to be no more than 4 feet. Past that it, too, became "lazy" and would crawl no more. Unfortunately, this lesson had to be learned the hard way: the commission had recommended digging shallow holes less than 6 feet deep in an earlier attempt at controlling the spread of the parasite, but this practice actually aided and abetted its ability to infect unsuspecting individuals who innocently answered nature's call, only to wind up infected with a parasite.

Over the next ten years, trial-and-error drug tests determined that orally administered thymol was well tolerated by children and adults alike, further reducing the community worm burden. That finding led to a treatment strategy that was coupled with programs aimed at educating the public on the need to sanitize. These two measures contributed in a major way to the decline in hookworm disease but did not reduce the prevalence of the infection. It was the relentless post–Civil War surge of the southern white population, abandoning their life as cotton farmers and settling in cities, that sociologists now believe led to the dramatic decline of iron-deficiency anemia caused by hookworm in the South. Today there still remain pockets of endemic hookworm, mostly in Appalachia and the deep rural South (Mississippi, Louisiana, Alabama, southern Georgia). Given that it was likely population changes that quashed the hookworm epidemic, the importance of the RSCEHD in altering the overall pattern of this disease during the commission's fifteen years of service remains questionable.

That said, the RSCEHD made some important discoveries. For instance, a seminal finding of the hookworm commission was its discovery of the relationship between the distribution of the infection and soil types. Those people living on hard-packed clay soils were largely parasite free, while those working plantations on sandy, loamy soil types were most affected. The explanation for these observations relates to the survivability of the infective (filariform) larva, which crawls in the ground and later penetrates into skin to initiate the infection. Moisture is critical to keeping the immature worm alive long enough for it to have a chance of encountering a host. Although the parasites lose their ability to migrate horizontally in soil after crawling 4 feet, they can survive at that spot for up to a month, given the proper environment. Each day they crawl up out of the soil in the morning to take advantage of the dew on the blades of grass. There they begin questing for a host. As the sun comes out and dries up the dew, the larvae migrate back down into the soil until the next day. Clay-bearing soils are more compacted and thus can hold far less moisture than sandy/loamy ones. That is why clay soils are unsuitable for the long-term maintenance of the immature hookworms. Two fledgling institutions in the United States got a real boost from these findings: the U.S. Public Health Service (1912), and the U.S. Geological Survey (1878). In particular, discovering that soil types could predict the distribution of a disease was the impetus for the USGS to commit to classifying all soil types in the United States.

The Rockefeller Sanitary Commission also set a valuable precedent for public health initiatives in the United States. The modern version of the public health service gained ground once the commission established guidelines for intervention with infectious diseases. It began issuing publications on a wide variety of health-related topics, including proper design and execution of the outhouse, largely based on the life cycle of the hookworm. In doing so, the outhouse soon became an icon for rural life in America. Modern composting toilets owe their origin to that humble structure. Along with hookworm, widespread acceptance and routine use of the outhouse as a way of controlling contamination of the environment with human waste also prevented the transmission of a wide variety of other, unrelated infectious disease agents: all those causing dysentery and diarrhea—polio, rotavirus, cholera, salmonella, giardia, entamoeba—and a number of soil-transmitted parasitic worms (e.g., ascaris, trichuris, strongyloides).

In 1913 a new entity, the Rockefeller Foundation, was granted nonprofit status by the New York State legislature. The next year it funded several international health initiatives, including a more global version of the hookworm commission, the International Health Board. The IHB's original charter included partnering with the Peking Union Medical College in China and focused squarely on eradicating hookworm from the Chinese mainland, an ambitious program to say the least. The IHB spun off another scientific body, the China Hookworm

Commission (CHC), which consisted of a new batch of American and Chinese hookworm biologists, some of whom were destined to become legends in the annals of American parasitology: William Walter Cort, Norman Stoll, and J. B. Grant. Stoll was already an accomplished young scientist who specialized in hookworm biology. I had the honor of knowing him during his last few years as emeritus professor at Rockefeller University, when I was a guest investigator there in the 1960s. He would often show up for our brown-bag lunches held in the conference room of another legendary parasitologist famous for his work on malaria, Dr. William Trager, and spin a story or two about his experiences with the IHB. As we sat quietly, all eyes transfixed on the great scientist's weathered, glowingly friendly face, he would proceed to hold court. Stoll was a master wordsmith and a most entertaining raconteur. At best, I can only paraphrase his brilliant use of language and humor.

Stoll recounted:

> Upon arriving in China, we began by sharing the early successes of the RMCEHD with the Chinese professors who had invited us, particularly with respect to sanitation. We thought all we had to do was to sit them down and explain how we figured out how to get rid of hookworm. If necessary, we were prepared to draw pictures of what an outhouse was and how to construct them. But we immediately encountered an unexpected problem when it was learned that the average Chinese farmer relied heavily on the use of human feces as fertilizer. Installing outhouses was not an option for controlling hookworm, so what were we supposed to do? We reevaluated the situation and began a series of experiments using whatever was available, in this case large vats the Chinese called *kongs*. These well-crafted and often highly decorated ceramic vessels were some two to three feet wide at the mouth and perhaps as much as three feet deep. We filled them with urine and feces, covered that odoriferous mélange with water, and eventually determined that if the kongs were left alone

for a minimum of two weeks, then that potent mix created an environment that killed the hookworm eggs before they could develop into larvae. Mixing in lime powder hastened the killing process and reduced the killing time from weeks to days. Constructing wooden sheds inside of which were housed a row of kongs supplied the farmers with a ready means of collecting their own wastes and also allowed them to use the "aged product" for fertilizer.

While the CHC had finally found a way to get rid of hookworm without relying on the outhouse to do so, their highly creative method did not allow the Chinese consumer of those crops the option of avoiding a plethora of other fecally transmitted bacterial and viral pathogens that survived quite nicely in the kongs for months. Schistosome, ascaris, and trichuris eggs and cysts of giardia, cryptosporidium, and entamoeba, and a wide variety of viruses including polio survived too and continued to plague farmers and consumers alike. These health risks still exist in many parts of China and Southeast Asia. Today the World Health Organization estimates that over half the world continues to employ human feces and urine as their fertilizer of choice.

Before melting down a few hookworm swords and recasting them into our plowshares, we need a better understanding as to what makes this parasite a parasite. Hookworms fall into the general category of nematodes. There are three known species of hookworm that routinely infect humans: *Ancylostoma duodenale, Necator americanus,* and, more uncommonly, *Ancylostoma celanicum.* It's even possible for people to be infected with the dog hookworm, *Ancylostoma caninum* (fig. 2.1). All hookworms share a similar life cycle, depicted schematically in figure 2.2.

The infectious larva begins its journey to our small intestine by penetrating the unbroken skin, typically through a hair follicle. To do so, it deploys at least one secreted protein in the metalloprotease family of proteolytic enzymes, allowing it to literally digest its way into the

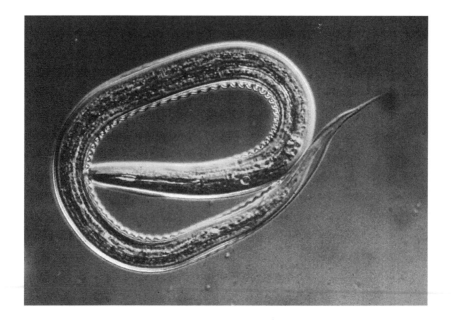

Figure 2.1 Infective larva of *Ancylostoma caninum*

bloodstream once it reaches the bottom of the follicle. As you will recall, this is the part of the worm's infectious process accidentally elucidated by Arthur Looss. Once in our circulation, the diminutive worm is carried passively throughout the body until it gets to a lung capillary. There it gets stuck, as its diameter is too large to fit through that narrow tube. Feeling the pressure of being squeezed against the capillary's endothelial cells, the immature hookworm reacts (a phenomenon known as *thigmotaxsis*) by penetrating out of the blood vessel. This mechanism is similar to the one that induces the migrating larva of trichinella to leave a capillary and penetrate into a muscle cell (see fig. 1.3). However, while trichinella uses a spearlike structure in its oral cavity to mechanically do the job, the hookworm larva does not possess such a handy little tool. To exit the lung capillary, it has to once again employ a metalloprotease

Necator americanus

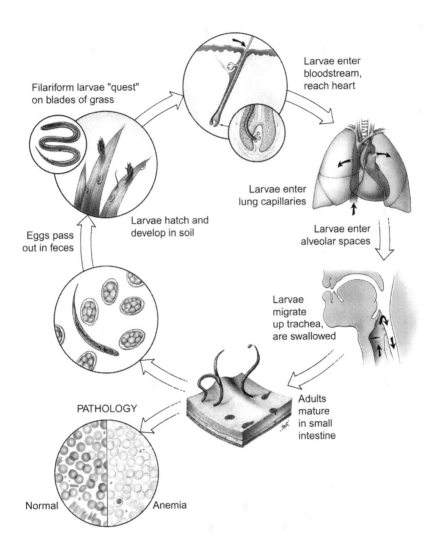

Figure 2.2 Life cycle of *Necator americanus*. Illustration by John Karapelou. (*Source: Parasitic Diseases*, 5th ed., (c) Apple Tree Productions, LLC, New York)

protein, similar to the one that facilitated its entry into the body at the start of the infectious process.

The hookworm larva is now out of the bloodstream, lying within the alveolar space of the respiratory tract. It then begins its long crawl up the bronchial tree until it eventually reaches the epiglottis, the anatomic junction between the lungs and the esophagus. When an infected individual swallows, waves of muscular contractions, known as peristalsis, carry the larva first to the stomach and then into the small intestine, where it will live for the rest of its life. Home at last!

As a side note, some people living in many parts of Asia and Papua New Guinea have a reduced hookworm burden, despite the fact that they are constantly exposed to the skin-penetrating larvae, because they have the habit of chewing betel leaves and routinely spitting out the juice along with their saliva. Any larvae that crawl up and out of the epiglottis and into the oral cavity become expelled along with that juicy compote. This practice, unfortunately, is not without serious drawbacks. Chewing betel leaves over a sustained period of time (years) has been show to be carcinogenic. Not a healthy trade-off for avoiding an infection with a blood-sucking parasite.

Back in the small intestine, the immature parasites have already molted twice to transform into infective larvae. They mature to adults by shedding their cuticle two more times. All this developmental activity occurs within a week after they reach the duodenum. With each new molt, the developing worms become more adultlike, building either male or female reproductive organs, mating appendages, and such. The sexes are distinctly different in shape and size, but they both have a similarly structured oral cavity that takes on a truly menacing appearance with the growth of cutting plates, or teeth (each hookworm species uses a slightly different arrangement of oral weaponry; see figs. 2.3 and 2.4). Mature worms attach to the villous tissue using their newly acquired dentures and then get to work making more hookworms. Their first task in this regard is selecting a mate, and once they do, the pair

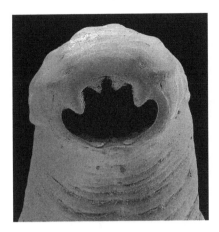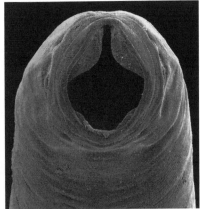

Figure 2.3 (*left*) Mouth of *Ancylostoma duodenale*. Photo by D. Scharf.

Figure 2.4 Mouth of *Necator americanus*. Photo by D. Scharf.

remains *in copula* for some three to five years. Not a bad existence for a couple of low-lifes! Their food source is the fingerlike projections of the outer surface of the small intestine (i.e., the villous). Each worm in the pair feeds by biting off a piece of intestinal tissue and digesting it in its tubular gut tract, employing proteins called cathepsin B cysteine proteases. It is thought that most of the worm's nutrition comes from this single source, but hemoglobin also contributes to its protein intake. Regardless of the source, most of the amino acids and other metabolites the parasite obtains from the host are used to make eggs and sperm. Females produce around fifteen thousand ova each day (fig. 2.5), but actual numbers vary according to each species. For an infection lasting on average four years, each adult female hookworm could end up producing as many as 217 million eggs.

Every villous also includes a capillary and a lymphatic vessel, and the adults take advantage of that by pumping blood emanating from the severed blood vessel through their gut track, making use of a

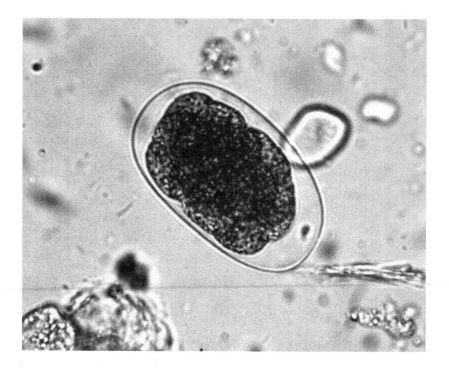

Figure 2.5 Hookworm egg

well-developed muscular esophageal bulb to do the heavy lifting. It is believed that this process allows hookworms to extract oxygen from our red cells. After they have pumped the blood through their gut tract, it flows out into the lumen of our small intestine. It is as though the worms had become an extension of our own circulatory system—albeit an open-ended one. It takes them about three to five days to completely devour a villous and move on to the next one. When they relocate, the abandoned site continues to bleed for another two to three days, until the host can repair the damage.

The continued leakage of blood into the lumen of the small intestine adds insult to injury with respect to our ability to conserve iron, a vital component for red cell production. If enough worms are present, over time this aspect of the infection can lead to iron-deficiency anemia, and this is the one activity of heavy hookworm infection that creates all the clinical problems for the host. It was the underlying cause of "southern laziness," and it's even possible that it could have quite literally sucked the fighting spirit right out of Johnny Reb. One recent hypothesis suggests that the Civil War was destined to favor the North as the winner from the start owing to the fact that most of the Confederate army was, at one time or another, suffering from anemia caused by hookworm—a parasite that southerners themselves brought to America as a hitchhiker inside African slaves. What an ironic piece of history that would be, if proven to be true. Not even the likes of such gifted writers as Nathaniel Hawthorne, Herman Melville, William Sidney Porter, or Rod Serling could have written a more twisted, moralistic, self-fulfilling prophecy. It must be quickly added, however, that hard scientific evidence favoring this scenario would be difficult to come by, unless we were to somehow develop a set of remarkably sensitive DNA-based tests for anemia and for the presence or absence of hookworm infection that could then be performed on the remains of both the Blue and Gray soldiers.

To keep our blood flowing freely, while at the same time feasting ad libitum on hunks of raw meat (a rather unsettling and obviously disgusting thought), the adult worms employ a single peptide, a small, protein-like molecule that inhibits the clotting of human blood. This unique parasite peptide is synthesized, stored, and secreted by two specialized esophageal glands, each equipped with a duct leading directly to the anteriormost part of the parasite's oral cavity. No other blood-sucking organism has evolved to this high a level at any time during the evolutionary history of anticoagulants. Thus it is by unanimous proclamation by the animal kingdom that the hookworm family be elevated to the collective presidency of Mother Nature's elite Go With The Flow Club. Mosquitoes and other biting flies, kissing bugs, fleas, ticks, mites, leeches, and vampire bats all pale by comparison (pun intended). The only one that comes close to hookworm adults regarding the length of time it clings to its host is the lamprey eel. All other bloodsuckers are ordinary card-carrying members compared with these amazing parasitic worms, blocking the cascade at only one of its many steps. I call them eat-and-run parasites. Their job is to latch on to their hosts, get what they need, and get out, sometimes in just a matter of seconds (mosquitoes, biting midges, and blackflies). Having a two-pronged attack in place that prevents clotting guarantees that hookworm adults can eat, drink, and be merry 24/7. So now we have a partial explanation as to why this is one of my favorite long-lived parasites.

Hookworms inhibit the clotting of blood for two reasons: first, so that they can feed on host tissue without having the blood they also ingest clog up their oral cavity. If they did not produce an anticoagulant, they might not be able to swallow hunks of our intestinal tissue. Second, it is believed that the adult worms derive oxygen from the blood they continuously ingest, so clotting would also prevent this essential function. It's likely that the worms would die if they could not access the oxygen in our blood as they feed. So producing the anticoagulant is an essential part of their biology.

In any given population at risk, hookworm disease disproportionately affects children, causing adverse long-term effects: permanent learning deficit, and an endocrinopathy characterized by an overall retardation of growth and development. Dr. Peter Hotez, a widely respected pediatrician who has spent most of his professional life studying various aspects of hookworm, maintains that this infectious agent alone is the leading cause of "failure to thrive syndrome" endocrinopathy worldwide. The mechanism by which this worm effects such global changes in our metabolism during the early phase of our growth cycle remains a complete mystery. By the time a chronically infected person achieves adulthood, most of the adult worms have been expelled by the slow, steady onset of acquired immunity. At the same time, although exposure rates to the infective larvae are about the same in grownups as they are in children, repeated reinfection in adults results in fewer and fewer worms maturing to adults. While most epidemiologists consider these findings in the field scant evidence, we can still draw the conclusion from them that there is real hope for developing a vaccine that elicits immunity against the skin-penetrating stage and the adult parasites. There are some promising leads in this regard, and at least one vaccine trial is currently under way in China, led by Hotez and his colleagues.

While the precise details remain unclear as to how acquired immunity against hookworm infection works, some general characteristics have been gleaned from laboratory and field investigations. We know, for example, that the life span of the adult parasite is limited by a coordinated set of humoral immune responses of the TH_2 variety. Details of its mechanism(s) are now beginning to be deciphered, but its still unclear how it develops and what part of the parasite (e.g., its digestive enzymes or the peptide anticoagulant) it targets. Furthermore, while our immune system also limits the degree to which we become reinfected, the precise mechanisms (antibodies or immune reactive cells such as macrophages) that target the larval parasite are still under investigation. Understanding all this will help to explain why hookworms may hold the key to unlocking the reason(s) that some of us become allergic to certain foods such as wheat gluten (celiac disease), while others go on to develop even more serious illnesses, like Crohn's disease, asthma, and multiple sclerosis, since populations living with hookworm infection have reduced rates of these allergic conditions.

Plowshare Concepts

We have several classes of hookworm-specific compounds to choose from to enrich our own pharmacopeia: those that we could apply to a vaccine to prevent hookworm infection, and others that we could apply to areas of medicine unrelated to the infection process (treating food allergies and certain autoimmune diseases). Vaccines could be deployed in areas of the world where intervention of infectious diseases by sanitation is difficult to achieve. The first group of parasite products that may prove useful in developing a vaccine includes the ones the immature parasite secretes to penetrate our skin, the metalloproteases. The goal of an effective vaccine would be to prevent infection by eliciting protective immunity against the infective larvae as they attempt to penetrate our skin. This kind of immunization would be particularly beneficial for very young children living in endemic areas. The second molecule of interest for vaccine development is the anticoagulant peptide produced by the esophageal glands of the adult worm. Immunity against this molecule would probably result in starving the worm, since it would be unable to feed for any significant length of time before it weakened and was expelled by the host. The third group of parasite products is the proteases, which the hookworms use to digest our hemoglobin. Again, if such a vaccine targeting these digestive enzymes were to prove effective, its mechanism would be to starve the adult parasites.

The plowshare concept, however, is what we can learn from the hookworm's powerful anticoagulant abilities. The general medical condition known as hypercoagulopathy includes diseases or conditions that result in a shortening of the clotting time of blood (clinically referred to as an embolism) and may rapidly escalate into to a life-threatening situation. These conditions include sitting or lying in bed for prolonged periods of time, any surgery, most cancers, pregnancy, and use of birth control pills. The need for developing new anticoagulants is a high priority for many pharmaceutical companies, and some have turned to the medical leech, *Hirudo medicinalis*, for inspiration in developing the next generation of compounds. But why focus on an organism that can block clotting in just a single place along the cascade when hookworm offers us a much more efficient way of dealing with those situations? Pathogenic microbes have adapted to generations of antibiotics and are now resistant to most of them. Newer approaches for drug development are required if we are to stay ahead of their evolutionary curve. At the same time, we have the notion that as we learn more and more about how our genome works, we can develop cures for inherited diseases through genetic manipulation. Assuming we succeed, this will still leave a large number of illnesses for which we have no answers. For those, we will have to use all the resources that nature has provided us through millions of years of evolution. In this regard, parasites rule!

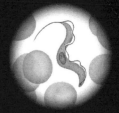

3 HOUDINI'S NEFARIOUS COUSINS

The Trypanosomes, the Schistosomes,
and the Lymphatic Filariae

Long-lived parasites have somehow, despite our best efforts, learned
how to thumb their noses at our immune system, escaping detection
by a wide variety of mechanisms. I have already documented the lives
of two long-lived groups of nematode parasites, *Trichinella spiralis* and
the hookworms. Here are some more, each employing a completely different mechanism to achieve the same goal: a long and productive life
at our expense. Some of them crawl into places we cannot reach with
our potent antibodies and activated immune cells, while others cover
themselves with our own serum proteins and pretend they are us. A
few don't even seem to mind being knocked off by the millions, as long
as one or two individuals escape. Collectively, they are indeed the rock
stars of the parasite world. These uberpathogens rose to prominence as
the result of millions of years of selective pressure, during which time
those that did not have the proper set of genes needed for an extended
life in their hosts were evicted by immune responses. So without further
ado, let's take a peek into their secretive worlds and learn some things
about the tricks they have evolved.

The African Trypanosomes: *Trypanosoma brucei rhodesiense* and *Trypanosoma brucei gambiense*

Trypanosomes are single-celled organisms (protozoans) that move by using a whiplike structure called a flagellum. They infect humans and a variety of wild animals. In Africa, trypanosomes cause a disease known as sleeping sickness. Different species of trypanosomes are found in the Americas, and one of them causes a deadly chronic disease called Chagas' disease. The majority of trypanosome species are transmitted to their hosts via the bite of infected insects. In Africa tsetse flies transmit the infection, while in the Americas it's kissing bugs that do the job, and there are many species of each insect group that play important roles in transmission in both geographic regions. The African trypanosomes spend their entire lives swimming about in their host's bloodstream, while ones in the Western Hemisphere spend most of their lives inside the host's cells. The African trypanosome group is the archetype for how to outsmart the antibody arm of the immune system and keep on ticking. This is achieved through the molecular equivalent to a combination of a magician's use of smoke and mirrors—now you see them, now you don't—and a quick-change artist's techniques. Just how they pull off this stunt is fascinating biology. New World trypanosomes employ a totally different set of strategies for escaping, essentially busting out of our cellular equivalent of jail.

From an evolutionary perspective, both trypanosome groups must have at one point been quite similar, if not identical, prior to the drifting apart of Africa and South America some 150 million years ago. Then, as the continents separated over the next 50 million years, the trypanosomes drifted apart in their biology as well. Today we see the result of their geographic isolation and speciation.

The discovery of African trypanosomes as causative agents of disease in cattle and humans occurred over a period of about twenty years, during the early phase of the golden age of parasitology, 1890–1910. Africa was largely unexplored (at least by Europeans) at that time, so many a would-be parasitologist set out for that continent in search of new infectious diseases. It should be pointed out that searching for medical fame in places that had never been visited before was not without its risks: death from accidents, infectious diseases, attack from venomous or predatory animals, and sometimes hostile behavior from a few very unfriendly indigenous human populations were commonplace. Despite all the potential hazards, amazingly, many a microbe hunter persevered and struck pay dirt.

David Bruce was one such individual. A surgeon by training and enlisted in the British Army Medical Service, Bruce began his career on the island of Malta, where he speculated that spontaneous abortion in cattle might be caused by a microbe. Shortly thereafter it was indeed shown to be due to a bacterium, and the genus Brucella was chosen to honor Bruce's contributions to the epidemiology of that disease. Brucellosis remains a serious infection of cattle to this day in many parts of the world.

Bruce then moved on to Africa and began a new series of epidemiological investigations that in 1903 resulted in his discovery of trypanosome infections in cattle. The disease it caused was known as *nagana* by the local herdsmen. Bruce also discovered that it was transmitted by the bite of the tsetse fly. In 1902 R. M. Forde had already described a similar infection in a patient from West Africa (Gambia) in whom the organism apparently had induced coma, then death. Forde, together with J. E. Dutton, directly observed the trypanosome (fig. 3.1) in the blood of that patient, making the connection between *nagana* and the disease they came to refer to as "sleeping sickness." Although Bruce looked extensively in other animals in West Africa (e.g., baboon, leopard, antelope, elephant), he could not find evidence of a similar trypanosome infection in them. Further investigation showed that trypanosomes

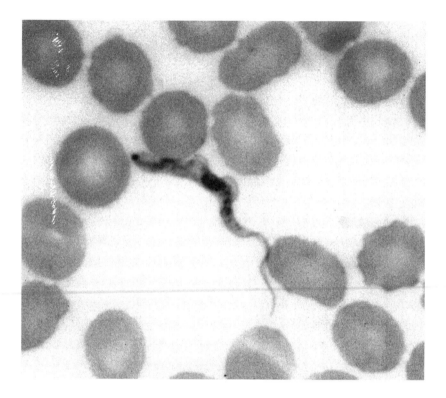

Figure 3.1 Trypomastigote of *Trypanosoma brucei*

caused disease in East Africa as well, but only in people. Like the West African version, it was transmitted by the bite of an infected tsetse fly. But unlike in West Africa, the bloodstream of the large mammal fauna of East Africa (e.g., lion, cheetah, wart hog, wildebeest) was teeming with parasites. Remarkably, none of these wild animals exhibited any of the clinical symptoms of *nagana*. In fact, they all seemed to tolerate trypanosomes as they would any other harmless microbial hitchhiker. And that wasn't the only puzzle. When Europeans established large ranching operations in East Africa at the turn of the nineteenth century and tried to establish herds of cattle of various types imported from their northern homelands, no introduced bovine fared well; all suffered from *nagana* and eventually expired. Bruce spent the next twenty years of his career trying to resolve the differences between the West and East African trypanosomes, but to no avail. One of the points of confusion lay in the fact that both trypanosome groups appeared to be identical in morphology when observed under the microscope, no matter what stains were used to make them more visible.

We know a great deal more about these two parasite populations and their biological properties since Bruce ended his research on them, now that their entire genomes have been completely sequenced. As Bruce suspected all along, they are indeed two distinct species: *Trypanosoma brucei gambiense* and *Trypanosoma brucei rhodesiense*. While they share many common characteristics, each has a unique set of traits that distinguish it from the other. But it is their shared molecular escape strategy that continues to hold the interest of parasitologists.

Before discussing the actual molecular mechanism that these protozoans have evolved, a brief review of our immune system is in order. Under the great majority of circumstances, the outer exposed surface of an invading microbe contains unique compounds that are easily recognized by our immune system. Upon encountering them, we respond by generating antibodies that specifically combine only with those compounds in a lock-and-key fashion. This usually results in neutralization of the microbe, and we recover from the infection and resume our normal lives. Alternatively, we can attack the microbe all out with our defense cells (e.g., macrophages, neutrophils, eosinophils), typically killing it by a process known as phagocytosis. Often we are not even aware that we have beaten off the enemy within. In the case of antibody production in response to an invader, we acquire the ability to mount an even stronger and quicker response the next time one of its kind gets the idea to parasitize us. That is why live vaccines of some viruses work so well.

There are, however, some remarkable microbes that have figured out a way to surmount both lines of defense, and the African trypanosomes are the best known of those that fall into that special category.

The African trypanosomes stand alone among all the organisms of the world regarding the mechanism by which they persist in their mammalian hosts. The surface of the bloodstream form is completely covered with a special protective coat of protein one molecule thick. Because the coat protein is decorated with various types of sugars, the general group of compounds to which it belongs is referred to as glycoproteins. Of the 9,068 genes in *Trypanosoma brucei*'s genome, 806 of them are committed to encoding these special proteins. Each gene codes for a single protein, called a VSG—a variant surface glycoprotein. Thus, coding for VSGs is nearly 10 percent of the total genomic investment by that organism. It is through the sequential deployment of VSGs that the organism engineers its survival in the host, despite a robust immune response against each variant.

In contrast, humans have approximately 27,000 genes, and close to 200 of them are dedicated to encoding antibodies of several major types. Yet amazingly, we have the ability to respond specifically to nearly a million different foreign compounds (antigens). If we were to have a separate gene for each of these specific antibodies, as the trypanosomes have for each VSG, then our genomes would be, by far, the largest on the planet. Just storing over a million different genes in our chromosomes would present quite a problem, let alone sorting them out for expression during an infection. Clearly, another mechanism is at work, enabling us to survive repeated infections from a wide diversity of would-be pathogens over our 200,000-year history on planet Earth.

The answer lies in our cells' ability to combine portions of the antibody molecule. The specificity of each antibody response to infection is generated by a special recombination process called *somatic mutation*. The required repertoire of antibody molecules is selected for when our antibody-synthesizing cells are exposed to an antigen. Upon contacting the antigen, only the cells capable of making the needed antibody (i.e., the one that combines with the antigen) are stimulated to multiply. This increases the concentration of the correct antibody type in the serum until the offending antigen is fully neutralized, after which the immune response wanes. The next time we encounter the same antigen, the response occurs sooner and with greater intensity because we generate long-lived cells during the initial immunization process that stick around and remember that specific encounter. These are called memory cells. That is why vaccines in childhood can protect us through the better part of our adulthood. In this way we have evolved a means of generating a wide range of protective antibody responses without the need for creating a separate gene for each one.

While trypanosomes are decidedly simpler than we are in their molecular organization, they are still able to elude our sophisticated defenses using the sheer numbers approach, coupled with a bait-and-switch strategy. To know how all this works, it is important to be able to visualize the actual arrangement of the VSG on the parasite. Although the amino acid sequences of VSGs vary widely, they all have common gross molecular features that allow them to fold similarly. In fact, different VSGs are almost indistinguishable when their three-dimensional crystalline structures are compared. Coat proteins are cylindrical in shape and line up on end when arranged along the membrane surface of the parasite, sealing off almost every part of it from immune attack by antibodies, save for a small, nearly inaccessible indented pocket through which its flagellum emerges, and through which it also absorbs nutrients from the host's bloodstream. One end of the VSG is firmly attached to the trypanosome's cell membrane by a special structure called a glycolipid anchor, while the other end, decorated with sugars—the antigenic portion—is exposed to our immune system.

We can make antibodies only against the exposed region of the VSG, but that is enough to eventually kill off every organism sharing this antigenic signature. We become infected after the bite of an infected tsetse fly, and the trypanosomes begin their relentless multiplication scheme in the bloodstream (fig. 3.2). Within a week or two, their populations have swelled to the millions. But in comes the cavalry: antibodies to the rescue! As our immune system catches on to the presence of the invaders, a vigorous antibody response ensues, eventually killing millions and millions of trypanosomes. If that was all that occurred, then the infection would be cured. But the trypanosmes have only begun to fight back.

Remarkably, in the time it takes our body to mount an antibody-based defense, a few trypanosomes will have already shed their first coat of VSGs and replaced it with a new one made of a different antigenic signature, VSG 2. No problem, our immune system says, as it ramps up for another round of antibody responses. Again, millions of trypanosomes now sporting new coats of VSG 2 die. I think you can guess what's going to happen next. Yet another few trypanosomes have already made the switch to a coat made exclusively of VSG 3. They then grow up to replace their fallen comrades, and the host once again has to regroup and make a new antibody response. It can go on like this for some time, even months. But eventually the buildup of cellular debris from all those dead protozoans, the constant depletion of blood glucose levels by reproducing populations of trypanosomes, and the buildup of their metabolic by-products takes its toll on our constitution. The trypanosomes then enter our cerebrospinal fluid and administer the coup de grâce. The host falls into a coma. In the case of West African trypanosomiasis, the coma comes slowly, while the East African version of sleeping sickness develops soon after infection and is fulminating. Without intervention using chemotherapeutic agents, trypanosomiasis is universally fatal. Those in East Africa die from the direct pathological effects of the protozoan, while those in West Africa usually die from pneumonia or some other secondary infection acquired over weeks to months of lying in coma.

One of the more interesting features of the VSG system is the fact that the parasites express them in a predictable order, starting with VSG 1, then 2, and so forth. In experimental infections in animals, if

Trypanosoma brucei gambiense and *T. b. rhodesiense*

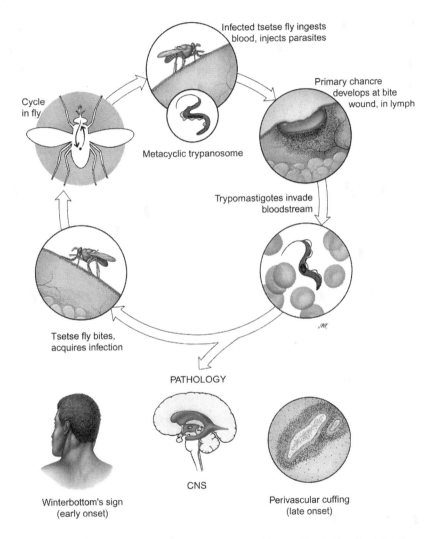

Figure 3.2 Life cycle of *Trypanosoma brucei gambiense*. Illustration by John Karapelou. (*Source: Parasitic Diseases*, 5th ed., (c) Apple Tree Productions, LLC, New York)

the trypanosomes are allowed to undergo several cycles of VSG expression and then are placed back into a naïve animal, the parasites revert back to the original VSG coat protein and start over again. With 806 different VSGs, the host never survives long enough to allow for the expression of all of them. So the evolutionary value of the VSG system is to allow the organism enough time to become transmitted from the infected host to an uninfected tsetse fly. Humans play little or no role in the transmission cycle since there are far more infected animal species that harbor the infection than humans, especially in East Africa.

Fortunately, the clinical situation is not as hopeless as it may seem. There are several drugs now approved by the World Health Organization (WHO) that are useful in the treatment of trypanosomiasis: pentamadine, suramin, nifurtimox, and eflornithine. The bad news is that many of them are associated with serious side effects, and they are all expensive. That is why the WHO has been designated as the main distributor, and it provides them free of charge.

The East African version of the disease has many reservoir hosts, so its control at that level is nearly impossible. The species of tsetse flies that live there are low in population density, so eliminating even a single fly might clear a square mile of grassland of the vector. In West Africa, because the density of tsetse flies is so intense along the banks of almost every river, avoidance of that narrow strip of land is the only reasonable approach to controlling the spread of the organism.

The American Trypanosome: *Trypanosoma cruzi*

Carlos Chagas is arguably Brazil's most renowned parasitologist. In fact, there is a disease named after him—Chagas' disease—coined by his close friend, Ozwald Cruz. To return the favor, Chagas named the causative agent of the disease *Trypanosoma cruzi*. We know it today as American trypanosomiasis. It is a chronic, deadly infection in humans, but well tolerated in a wide spectrum of wild and domestic animals throughout South and Central America, and even some parts of the southwestern United States. According to the World Wildlife Federation, the Cerrado of South America is home to more species of large mammals than anywhere else on our planet: "The largest savanna region in South America, this ecoregion also contains an amazing amount of biodiversity. Located throughout Brazil, Paraguay, and Bolivia, over 10,400 species of vascular plants are found, fifty of which are endemic. Fauna diversity is very high also with 180 species of reptiles, 113 of amphibians, 837 of birds and 195 of mammals." Remarkably, a portion of virtually every population of mammalian species found there is infected with *T. cruzi*. Even those mammals living deep in the Brazilian tropical forests harbor the parasite.

Like its distant African relatives, *T. cruzi* (fig. 3.3) is a vector-borne infection, but in this case the vector is a two-inch-long bug with such a penchant for blood that it makes Bela Lugosi look like a vegetarian. The "kissing bug" vectors are widely distributed, and there are at least three major species that play a key role in the transmission of Chagas' disease. Unlike other vectors, it is not the kissing bug bite itself that transmits the parasite to the host. Rather, to make room for another blood meal, the insect actually defecates a large portion of its gut contents onto the skin of its current victim. The infectious trypanosomes are in the bug's feces and can gain entrance into the host only if the animal being bitten rubs them into the bite wound or into a mucous membrane. As unlikely as this transmission mechanism may seem to be on first inspection, the parasite takes advantage of the fact that the bite of the kissing bug induces a local inflammation and an associated intense itching sensation. Scratching the site of the bite wound is normal behavior for any animal and is usually enough to move some trypanosomes inside the wound. Often that is all it takes to initiate infection.

Animals can also become infected by consuming the offending bugs (dogs and cats do this when they see the bugs feeding on them). More

Trypanosoma cruzi

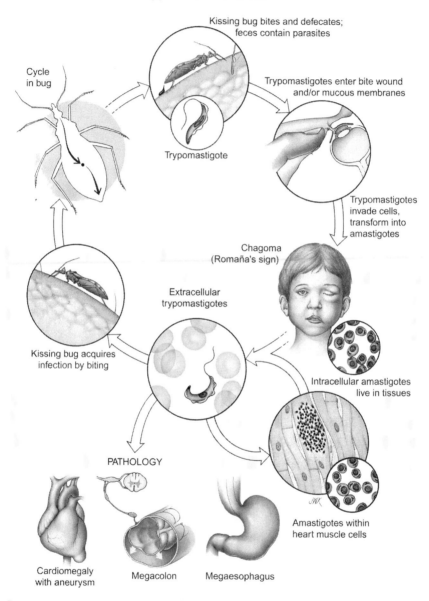

Kissing bug bites and defecates; feces contain parasites

Cycle in bug

Trypomastigote

Trypomastigotes enter bite wound and/or mucous membranes

Trypomastigotes invade cells, transform into amastigotes

Chagoma (Romaña's sign)

Extracellular trypomastigotes

Intracellular amastigotes live in tissues

Kissing bug acquires infection by biting

PATHOLOGY

Amastigotes within heart muscle cells

Cardiomegaly with aneurysm

Megacolon

Megaesophagus

Figure 3.3 Life cycle of *Trypanosoma cruzi*. Illustration by John Karapelou. (*Source: Parasitic Diseases*, 5th ed., (c) Apple Tree Productions, LLC, New York)

rarely, the infection can be sexually transmitted. Blood banks have been the source of transfusion infections in humans. One unusual transmission situation has arisen in recent years at the most popular beaches all along the Atlantic coastal areas of Brazil, including that country's most famous resort, Copacabana. It involves the manufacture of popular beach drinks made from either raw sugarcane or açai palm juice. When the plants are crushed to prepare beverages from them, any infected bugs included in the harvested plants also become pulverized, thus releasing their trypanosomes. Unsuspecting sunbathers then become infected by drinking these sweet concoctions. Children usually become infected at night while they sleep. Many of the rural houses of Brazil are typically made of thatch and wood. The bugs live in cracks in walls and in the roofs. Often an individual is bitten near the orbit of the eye. The term "kissing bug" refers to this situation, and to the fact that, despite their size, the bugs are rarely perceived by the host during the act of taking a blood meal because the insects inject an anesthetic that deadens the site of the bite wound without inhibiting the itching sensation that follows.

As detailed above, African trypanosomes live in the bloodstream and the cerebrospinal fluid of their hosts, alerting the host as to their presence by shedding their surface coat proteins. They do not avoid immune attack; rather they sequentially vary their antigenic signatures to continue the infection until the host becomes exhausted and eventually dies. In contrast, *T. cruzi* has evolved to become mostly a tissue parasite, inhabiting a wide variety of cells, including heart muscle, nerve cells, and even macrophages. The myenteric plexus of the nervous system (large collections of nerve cells) that enervates the smooth muscles of the small and large intestine is a favored infection site. Over time these nervous connections break down after repeated replication cycles of the parasites, resulting in a loss of smooth muscle function. This causes the hollow organs to become flaccid and renders them nonfunctional. The heart also enlarges, and death may ensue from cardiac insufficiency.

All this takes time. In fact, in humans, infections often go on for twenty or more years before severe pathological consequences develop. Treating the chronic phase of the infection involves the use of benznidazole or nifutimox.

Once the trypanosomes are introduced into the bite wound, some of them encounter—and are ingested by—dendritic cells. These large, macrophage-like cells, whose function was first described by Nobel laureate Ralph Steinman, live in our skin and are programmed to ingest any microbe that may stray into their territory. The vast majority of infectious agents are killed and partially digested so that their antigens can then be shared with other cells of our immune system, enabling us to make specific antibodies and protective cellular-based immune responses. This is how we defend ourselves against microbial pathogens. Ordinarily these dendritic cells would destroy the invader and that would be that. But *T. cruzi* is not your ordinary parasite. Since Chagas' disease can smolder for tens of years before its victim succumbs to the infection, how does the parasite manage to avoid destruction after it enters a dendritic cell, or any other cell for that matter?

The first thing *T. cruzi* does after being ingested by a cell is to round up and discard its flagellum. Then it secretes substances that inhibit the dendritic cell's molecular pump, which under most other situations would acidify the contents of the vacuole containing the parasite, leading to the death and digestion of the invading pathogen. Failure to acidify the vacuole allows *T. cruzi* to survive long enough to pull off its Houdini-like escape act. Before the dendritic cell can recover from its disarmed digestion mechanism, *T. cruzi* simply penetrates out of the membrane of the vacuole and insinuates itself into the naked cytoplasm of the dendritic cell. It is now hiding in plain sight; a most ingenious stealth move on the part of the parasite (fig. 3.4). But *T. cruzi* still has some host cell issues to deal with, despite its clever solution to the digestion problem.

It turns out that all mammalian cells have the capacity to kill invading microbes by activating an innate immune response system. This

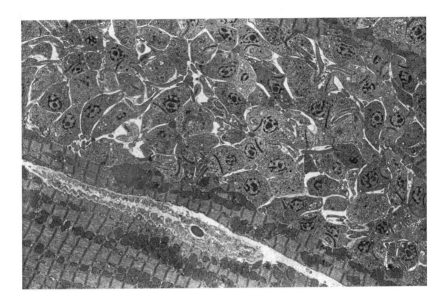

Figure 3.4 Epimastigotes of *T. cruzi* inside an infected cell. Electron micrograph by Charles Sterling.

requires a complex series of molecular events involving the translocation of a special protein transcription factor called NF-κB (nuclear factor kappa-light-chain-enhancer of activated B cells), which moves from the cytoplasm into the nucleus of the infected cell. As a direct result of NF-κB's interaction with specific sequences of our DNA, interferon and other related protective molecules are then synthesized, which act in concert to inhibit the multiplication cycle of the pathogen.

Remarkably, *T. cruzi* has evolved a survival strategy that takes care of this host response as well. Its surface is covered by a membrane into which is anchored a special protein, the 63 kDa cystiene protease, better known as *cruzain*. When the infected cell's NF-κB protein encounters the parasite membrane, a portion of NF-κB is digested by cruzain, rendering this important host protein inactive. The parasite survives and replicates. Chalk up one more victory for the parasite!

After dividing several times, *T. cruzi* overwhelms the infected cell and kills it, becoming freed into the spaces between the cells in the tissues, or directly into the bloodstream. The parasite then transforms back into its original elongated flagellated form and initiates infection in a new cell. Uninfected kissing bugs acquire the parasites by feeding on the infected host. This is why *Trypanosoma cruzi* is on my short list of superparasites. It seems to have everything going its way.

The Schistosomes: *Schistosoma mansoni, S. haematobium, S. japonicum,* and *S. mekongi*

In 1951 Humphrey Bogart and Katharine Hepburn made a wonderful movie, *The African Queen.* It was the only time they worked together, and it was a box office smash. It took place largely on Lake Victoria in Kenya and was set in time just before the outbreak of World War I. There were, of course, the memorable and endearing scenes of Bogart and Hepburn, bickering back and forth, as he manipulated his dilapidated vessel into position to eventually sink a German warship. Audiences the world over loved the movie for its sheer drama and excitement, although I bet most have long forgotten the details of that award-winning film. But there was one particular sequence that everyone who saw it can recall, and with all the gory details, too. It was when Bogart pulls the *African Queen* into position in an estuary of the lake. He doesn't want to alert the enemy, so he turns off the sputtering engine and gets into the water, shoulder deep, and begins the arduous task of hauling the ailing craft to the lakefront. When he's done pulling on the tow-rope, Bogart struggles back on deck, only to find himself covered from head to toe with big, black, blood-sucking leeches. He and Hepburn are thoroughly repelled by the sight of them, and she proceeds to rub them off of his back and legs by applying handfuls of salt to each of those "dirty little buggers."

Whew! In those days that was heavy stuff for a general audience. Nonetheless, I give the writers, director John Huston, and in particular Humphrey Bogart maximum credit for having the courage to show parasitism at work, up close and personal. Fortunately they had the good sense not to film those particular scenes on location in Africa, or they would have been in for a surprise of a different kind. In Lake Victoria another, even more sinister, parasite lurks in the water that could have done them all in: schistosomes. Throughout the tropics, these trematode parasites infect some 250 million people, and millions go on to suffer from a wide variety of illnesses caused by the chronic effects of their presence in the bloodstream. Remarkably, these worms can thrive inside us for as long as twenty to thirty years. All it takes to become infected is to wade or swim in freshwater aquatic environments in which the infectious stage is found, à la Bogart. The parasite does the rest by penetrating the unbroken skin, using its digestive enzymes, in much the same way that hookworm larvae gain entrance (see chapter 2). There are four distinct species of schistosomes, and all use the same strategies, described below (fig. 3.5).

The microscopic stage that swims about looking for us is called a cercaria. It seeks out its host in much the same way a heat-seeking missile searches the skies for enemy aircraft. Only instead of a heat signature, this parasite takes advantage of other environmental cues to detect our presence. To do so, it uses its well-developed nervous system as its radar device. The cercaria swims about looking to pick up even a faint trail of a chemical mixture consisting of fatty acids and L-arginine (an amino acid) that we produce in sebaceous glands within the skin. These chemical signals are released when we enter water, forming a thin film at the water's surface in all directions. The parasite can detect these host-specific substances even in very small amounts. Once it locks onto a victim, it speeds up its pace of swimming, always toward the highest concentration, until it makes contact with skin. Clever little beasts! The

Schistosoma mansoni

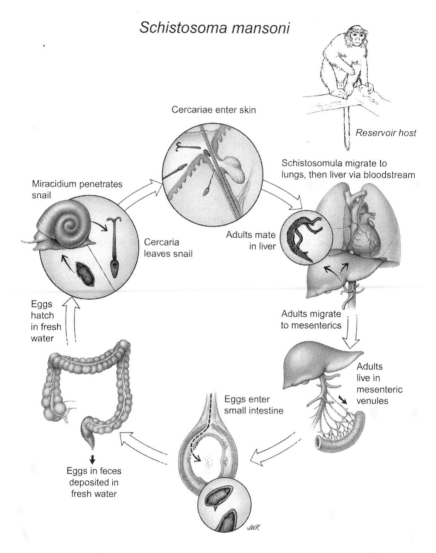

Reservoir host

Cercariae enter skin

Schistosomula migrate to
lungs, then liver via bloodstream

Miracidium penetrates
snail

Cercaria
leaves snail

Adults mate
in liver

Eggs
hatch
in fresh
water

Adults migrate
to mesenterics

Adults
live in
mesenteric
venules

Eggs enter
small intestine

Eggs in feces
deposited in
fresh water

JWK

Figure 3.5 Life cycle of *Schistosoma mansoni*. Illustration by John Karapelou.
(*Source: Parasitic Diseases*, 5th ed., (c) Apple Tree Productions, LLC, New York)

cercaria then attaches to us with its two suckers, discards the forked tail that it used for swimming, and proceeds to inch its way toward the nearest hair follicle. Upon contact with this entry point, the diminutive trematode squeezes down the shaft of the follicle—much like an octopus entering a vessel that looks too small to accommodate it—until it reaches the hair's root. This stimulates the parasite to secrete its collection of enzymes that digest proteins (called proteases), facilitating its entry into our dermis, the deepest layer of our skin. As improbable as this transmission scenario may seem from a biological perspective, the worldwide success of the schistosomes attests to the fact that they have evolved a very reliable infection strategy.

Having entered the dermis, the small, adolescent worm—now referred to as the schistosomulum—just sits there for up to three days. Scientists still don't know what it's doing during this time. Most likely it is adjusting its chemistry for life in a warm-blooded host after having just lived part of it as a free-swimming cercaria in freshwater. From the host's point of view, this is a critical stage of the life cycle since it has been well documented, both in the laboratory and in the field, that protective immune responses successfully target the schistosomulum in the skin. Microscopic evidence for this protective response shows that our eosinophils and certain classes of antibody molecules can act in concert to kill this immature stage of the parasite. Repeated infections are necessary to elicit these lethal dermal allergic reactions. Even though the parasites infecting a naïve host elicit protective immune responses, the host cannot make effective levels of protection quickly enough because the worms develop to the next stage too rapidly.

If the worm survives these three days in the dermis, the schistosomulum, after getting its metabolic act together, is ready to take the next leg of its journey, from our skin to the lungs. It does so by penetrating out of the dermis and into a small blood vessel. It is passively carried throughout the body by the circulation until it encounters the capillaries of the lungs. This stimulates the still-microscopic worm to penetrate out of the circulation and into the surrounding lung tissue. Then something fantastic happens: the worm turns on its cloaking device. It cannot be detected any longer by any branch of our immune system. From that point on, it can travel anywhere in the body without concern for waking up our immune system. It is home free!

How does it manage to pull off this incredible disappearing act? The schistosomulum has acquired the ability to incorporate foreign proteins—proteins native to the host—onto its surface, analogous to the way the decorated crab of the coral reef camouflages itself with sea anemones and coral polyps. Among those host proteins is a crucial molecule called β-2 microglobulin, found on the membrane of all nucleated cells. It plays a major role in our ability to respond to invaders. When a macrophage encounters a foreign object, be it a virus, bacteria, protozoan, or worm, it surveys the surface of the object for chemical hints as to its identity. If it fails to detect *self* molecules, such as β-2 microglobulin, it shifts into high gear and initiates a cascade of events eventually resulting in the destruction of the invader. Since the worm now has β-2 microglobulin on its surface, the macrophage surveys the worm and accepts the parasite as part of itself and moves on. Perhaps a brief description of a scene from another popular movie will reinforce this point.

The film is *Predator*. This highly successful sci-fi horror flick starred the governator, Arnold Schwarzenegger. The scene I want to use as metaphor for the cloaking behavior of schistosomes is the one in which Arnold is being chased through the jungle by a weird-looking alien who has come down to Earth for a bit of sport—hunting humans. Eventually Arnold ends up in a river and swims for his life. When he emerges from the river, he is thoroughly exhausted and can barely crawl up on the bank to face his attacker. And, he is covered from head to toe with mud. The alien follows him and eventually stands just feet away from his potential victim. Schwarzenegger expects to be killed and taunts the monster, but the predator cannot see or hear him because it relies solely on infrared signals to find its quarry. Arnold is now off the alien's heat detection system.

So it is with the schistosomes. In this case, the mud is our proteins (mostly soluble serum components), and many kinds have been found attached to the worm's surface (fig. 3.6). In fact, in laboratory experiments, when adult schistosomes are carefully transferred from one mammalian species to another—from mice to rats, for instance—the surface proteins on the parasites switch from the first host to the second one in a matter of days. In other words, the parasite, somehow sensing it is now in a different host species, sheds the surface serum proteins acquired from its first host and replaces all of them with serum proteins from the second host—a complete camouflage makeover. Furthermore, if the recipient host species is first immunized using normal serum from

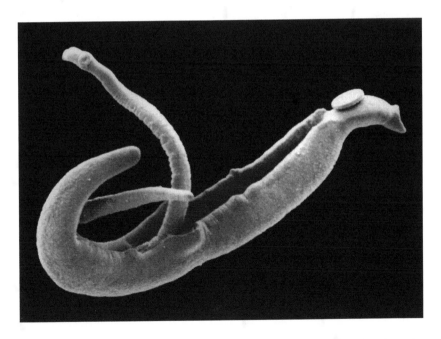

Figure 3.6 *S. mansoni* adult worms. Scanning electron micrograph. Photo by D. Scharf.

the first parasite donor host species prior to the transfer of adult schistosomes, those transferred parasites will be killed by the antibodies in the new host, showing how essential the cloaking device is in maintaining the life of the worm in an otherwise hostile environment. Recent experimental evidence suggests that the parasite molecule responsible for binding host serum proteins to its surface is called paramyosin, which is located in the outer layers of the parasite (tegument). Another study, however, suggests that a member of the tetraspanin family of proteins may be the molecule that does the job. In either case, when each of these proteins is used to immunize a susceptible host, the host is protected from a primary infection, showing that both may play a central role in the mechanism leading to the establishment of the stealth process. More studies are obviously needed to resolve this important issue.

From the lungs, the immature worms journey on to the liver via the bloodstream and feast on liver cells until they mature into adult parasites. During this phase, each individual worm occupies its own small domain within the liver tissue. It is good to keep in mind that the liver is the largest internal organ in the body. Remarkably, mature adult male and female worms, now 15–20 mm long, depending on the sex, are able to find each other in that vast organ by following trails of pheromones (chemical attractants) that are produced by both sexes. After migrating in liver tissue and encountering the opposite sex, the male embraces the female within his gynocophoric canal in a permanent act of copulation.

Up to this point, all four species of schistosomes behave similarly, but there are differences as to what happens next. In three species (*Schistosoma mansoni*, *S. mekongi*, and *S. japonicum*), the happy couples take the last leg of their fantastic voyage, moving as one against the flow of blood, out of the portal circulation and coming to rest in the mesenteric venules of the small intestine. In contrast, *S. haematobium* adults take a different route out of the liver, eventually occupying the venus plexus of the bladder. The maturation process in the liver (i.e., from immature worm to adult) takes place over several weeks, while the adults' life in the bloodstream afterwards can last up to an astounding twenty to thirty years.

Egg laying ensues almost immediately after the worms take up residence in their final destination. But there is a serious problem now facing the parasite. The eggs must somehow exit the host in order for the life cycle to continue, but they are produced within the lumen of the venules. There is no anatomic connection between the venules of the bloodstream and the lumen of the small intestine or bladder. So how do these fully embryonated eggs get to the outside?

The larva inside each ovum releases lytic enzymes that allow the egg to literally digest its way out of the host's body. This process causes damage to the small intestine and bladder. In terms of the clinical manifestations of schistosomiasis, it is the egg stage that does the most damage to the host. That is because only half ever succeed in escaping to the outside environment through the feces or urine. The rest wash back into the liver through the portal circulation and get stuck in the capillaries. As the liver's circulation becomes compromised over time by millions of eggs blocking the passage of blood, eggs become shunted to other organs like the brain, lungs, and pancreas. It is only then that we begin to exhibit the chronic pathological effects of infection with these long-lived, unwanted residents. This late-onset sequelae often proves fatal if left untreated. Eggs that do make it to the outside must enter freshwater for the cycle to continue. For most people or animals living in the tropics, a river or lake is a convenient place to urinate and defecate. This is also the perfect solution for the parasite; on entering the water column, the eggs hatch.

The freed larva, termed a miracidium, begins its own quest for a host, only in this case it's looking for a freshwater snail. Like its later stages, the ciliated, motile miracidium has a sensitive built-in radar system connected to its nervous system that it uses to lock onto the chemical trail of the snail. Once the snail is located, the larva attaches to the foot of its molluscan host and secretes proteases that facilitate its entry into the snail's hepatopancreas. The parasite undergoes yet another set of developmental changes, producing its infective stage for mammals, the cercaria. The infected snail can produce millions of these forked-tail swimming schistosomes through its own life span, which may last as long as two to three years. When a cercaria encounters and infects a mammalian host, the cycle is completed.

We have now come full circle in the life of the schistosome worm. Its life is filled with so many twists and turns that virtually no self-respecting biologist would have ever been able to conceive of such a fanciful, unbelievable scheme. It took hard work carried out over many years by dedicated parasitologists for the essentials of its life to be known. There is still much scientists have to learn, but with the advent of genomics and rapid DNA screening technologies, coupled with pro-teomics, it is only a matter of time before we will know what each and every gene does and when it is done. In the meantime, it is still pos-sible to imagine how we might use knowledge of how schistosomes hide from our immune system to our advantage, both to fend off the parasite and in other areas of human medicine.

The Lymphatic Filariae: *Wuchereria bancrofti* and *Brugia malayi*

Life might not seem worth living if every day you had to stare down both barrels of a fully loaded shotgun, but that is exactly what the lymph vessel–dwelling filarial nematodes, *Wuchereria bancrofti* and *Brugia malayi* (the causative agents of a condition commonly known as *elephantiasis*), do inside millions of unfortunate individuals who live in tropical settings all over the world. Swimming in lymph laden with immune cells seems like the very last place any self-respecting parasite would choose to live, but somehow they manage not only to survive there but to thrive in that unlikely environment. The molecular mechanisms responsible for this death-defying act are the subjects of this last section of *Houdini's Nefarious Cousins*. But before we begin our story, let's outline the general features of how they carry out their lives and try to define what we still need to learn about their lifestyles (fig. 3.7).

All lymphatic filarial parasites are transmitted by biting insects, and these two rely on mosquitoes (both culicine and anopheline species) to get from one human host to another. The geographic distribution of lymphatic filariasis (a strictly human infection), is currently restricted to the tropics and subtropics, but there was an endemic center established in the southeastern United States following the advent of the slave trade, beginning in the early 1800s. It was centered around Charleston, South Carolina, the entry port for enslaved West Africans. In fact, it was that very population of forced immigrants who inadvertently introduced the infection, along with hookworm and malaria, into the New World. Today there is no longer the threat of acquiring either of these filarial parasites in the United States, but nearby Haiti and the Dominican Republic still have remnant endemic vestiges of them. What is interesting about their introduction is that the New World already had competent insect vectors waiting for these parasites to arrive. Once the parasites were introduced, new transmission zones became rapidly established. This general process is still going on with other diseases; witness the introduction of the West Nile virus into North America in 1999.

While mosquitoes facilitate the development of the infective stage of the parasites and aid in their transmission, the microscopic worms are not actually injected into the host by the insect. Rather, they gain entrance into us much the same way as *Trypanosoma cruzi* enters the skin: through the bite site immediately after the vector obtains a blood meal. The infective larvae lie in wait in the lumen of the mouthparts of the female insect. When she begins to feed, the body heat from the host stimulates the parasites to move to the very end of the mosquito's hypostyle, and when she withdraws it from the bite site, the larvae crawl out

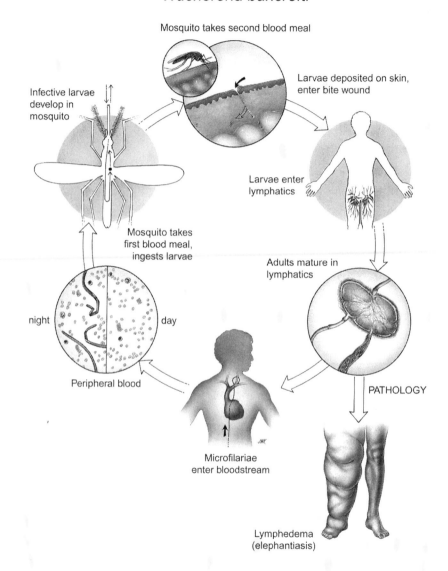

Wuchereria bancrofti

Mosquito takes second blood meal

Infective larvae develop in mosquito

Larvae deposited on skin, enter bite wound

Larvae enter lymphatics

Mosquito takes first blood meal, ingests larvae

Adults mature in lymphatics

night   day

Peripheral blood

PATHOLOGY

Microfilariae enter bloodstream

Lymphedema (elephantiasis)

Figure 3.7 Life cycle of *Wuchereria bancrofti*. Illustration by John Karapelou. (*Source: Parasitic Diseases*, 5th ed., (c) Apple Tree Productions, LLC, New York)

of the insect and onto the skin. From there they find the hole created by the mosquito's mouthparts and crawl into it.

The immature parasites then find their way via a capillary into the lymphatic drainage system and eventually reach a lymph node. They are unable to travel further owing to their large size. They complete their development within six months, maturing into threadlike adult males and females that measure between 4 and 10 cm in length. Soon thereafter they mate, and females give birth to live larvae, termed microfilariae, that measure some 220 μ in length. Because of their small diameter (about the same as a red blood cell), these immature worms, unlike their parents, can navigate their way past the lymph node and eventually enter the bloodstream. Microfilariae can live in blood for up to six months. When a female mosquito takes a blood meal from an infected individual, the microfilariae enter the stomach of the insect and immediately penetrate into the flight wing muscles. Within a week, the larvae mature to the infective stage for humans and migrate to the biting mouthparts. When the mosquito takes another blood meal, the larvae crawl out onto the skin and into the bite wound, beginning a new round of infection.

As mentioned, adult worms can live as long as ten years, but eventually they die. The result of the death of the worm is a blockage of the vessel caused by the cellular reactions to the dead worm material. Lymph then has to flow around the occluded vessel and find its way to the heart using another open lymph vessel. If enough worms are killed, complete blockage of the lymphatic drainage system in that area can occur, and lymph fluid begins to build up in the tissues behind the site of the blockade. Swelling ensues. Over a long period of time, perhaps months to several years, the swelling becomes so extensive that it results in gross distortion of that region of the body. In the lower extremities, the swelling is often so severe it resembles an elephant's foot, hence the term "elephantiasis." In other anatomic locations, the results are equally grotesque and disfiguring. In males, the scrotum can enlarge many times over (hydrocele), and in women, a breast may become enlarged. Often these swellings are so extensive that the individual is forced to stop working. Loss of sensation in the affected areas may result in inadvertent damage from trauma, secondary infection, or accidental severe burns. While drugs and specialized surgical procedures are available for treating these infections, most victims cannot afford any treatment. Mosquito control is an option, but with so many species involved in the transmission, vector control programs at that level are not effective and can come with their own problematic side effects, as with the indiscriminate use of DDT.

There is another aspect of the infection worth mentioning. It seems as though all species of filarial parasites harbor endosymbionts called Wolbachia. These are intracellular bacteria that live their entire lives inside certain cells within each stage of these highly evolved parasites. Their exact role(s) in the lives of these worms has yet to be determined, but eliminating the symbionts by treating the infected host with antibiotics often reduces or even ablates entirely the pathological consequences of lymphatic filariasis. The same is true for another related filarial nematode, *Onchocerca volvulus*, which causes blindness if left untreated. However, it is not thought that the endosymbionts play any role in their stealth mechanism(s).

Lymphatic-dwelling nematodes have evolved a unique strategy ensuring their survival: the secretions of the adult parasites induce a generalized hyporesponsiveness (i.e., a low-grade paralysis) of our entire immune system. During these five to ten years, there is little disease associated with the infection. But when the worms have lived out their lives, things begin to sour on both sides of the fence. The immune system that has been dozing wakes up and kicks into high gear to deliver the knockout punch that finishes the parasites off. Unfortunately, it is this unleashing of our cellular responses that does the damage to the lymphatic vessels. We end up hurting ourselves as well as the offending pathogens.

One intriguing feature of infection is the effect the adult worms have on the cells that comprise the lymphatic vessels (endothelial cells). The worm secretes growth factors that force these host cells to proliferate, which increases the diameter of the vessel. This facilitates the flow and keeps the parasites bathed in lymph throughout the rest of their life spans.

Plowshare Concepts

It is hard to envision how we might apply what we already know about the life of trypanosomes to the clinic. Their lives appear to be too specialized in directions not relevant to our needs. In contrast, the schistosomes and, to a lesser extent, the filariae offer the promise of helping to solve the host-versus-graft response that is so crucial in organ and tissue donation, as well as providing potential solutions to lymphatic drainage pathologies.

While it is well established that the adult stage of all four schistosome species is able to disguise itself using a "parasite in human's clothing" strategy, and successfully hide from our immune system, the exact molecular cloaking mechanism has yet to be discovered. Nonetheless, there are some intriguing clues regarding the identity of the parasite molecules that engineer this unique stealth strategy. In two separate laboratory studies, animals that were immunized with either adult worm–specific tetraspanin or paramyocin showed strong protective immunity after the animals were challenged with a primary infection. These results point to an immune-mediated disruption of the cloaking device, since only proteins from the adult stage were used. If these studies can be extended to show that tetraspanins and/or paramyocin are indeed the only proteins that are able to interact with the host's serum proteins, a strategy could be developed to enable tissue transplants between humans or even between dissimilar animal species without rejection.

Here is how it might play out, first in the laboratory and then at the bedside. Since all proteins are encoded by DNA, isolating and cloning the segment of the parasite's genome that encodes the cloaking protein(s), once their identities have been confirmed, would be straightforward. Then a series of pilot experiments would ensue as an initial proof of concept. These might involve employing rats and mice in experiments designed to prove that mouse organs can be transplanted into rats at will, without rejection, and can survive in the recipient host for long periods of time.

First, fertilized mouse eggs would be transfected with schistosome genes encoding either the tetraspanin or paramyocin proteins. After proving that both these parasite proteins were not deleterious to the growth and normal development of the transfected mouse, adult mice expressing either gene would then serve as the donor of various organs for rats. But before proceeding directly to the transplant part of the experimental protocol, it would be necessary to first demonstrate that both expressed parasite proteins localized to the cell surfaces of all organs of the donor mice, especially the endothelial cells that make up the bulk of the cells lining the circulatory system. If they did, then they would at least have the opportunity to interact with serum proteins from the recipient. Failure of recombinant parasite proteins to localize at the cell surface might be corrected by including a cell membrane-specific peptide into each parasite gene construct. I think we could eventually get this part to work without too much trouble. Next, we would have to show that any organ from these genetically altered mice could bind a wide variety of rat serum proteins in a random fashion. There are well-established immunological methods that could be employed for this part of the transplant journey to reveal the presence or absence of foreign proteins on the endothelial cell surfaces of the circulatory system of the donor mouse organs.

If all these preliminary studies proved out, we would then be ready for the transplant part of the experimental protocol. Transplanting any donor mouse organ into a rat must result in a "take" in order to go to the second experimental level. Ideally, the mouse organ should survive for the life span of the rat host without altering its life. This part of the work could take several years of repeated testing to document that nothing adverse occurred to the rats.

The next big step, if all went according to plan, would be to repeat the experiment using pigs as the transfected host species and successfully transplanting their organs into dogs, using the same methods as those used in the mouse/rat experimental trials. Adult pigs are about the same size as humans with respect to their organs and have been the subject of much research in the field of transplant surgery. Everything that was done in rodents must now be repeated. In addition, a strain of pig must be identified that does not pose a threat to humans with regard to the presence of virus infections that might also infect the recipients. There is a long list of these viral infections that must be ruled out before proceeding.

Success at this point would enable teams of transplant physicians to use these special pigs as donors to humans who are in a life-and-death struggle with illnesses that can only be addressed with a transplanted organ. Cirrhosis of the liver, numerous heart diseases including Chagas' disease, type 1 diabetes, Crohn's disease, and cancers (lung, pancreatic, liver) all fall into this category. I once presented a brief description of this scenario to my medical school class and ended my list of desirable pig organs with the whimsical possibility of pig brain transplants. I added quickly that, fortunately, most males already had one of those! The women in the class laughed, but, not surprisingly, the men were decidedly less amused.

A few filarial parasite compounds might also prove beneficial. The adult worm secretes soluble products into the lumen of the lymph vessels almost immediately upon arriving there. Some of them cause the vessel to dilate. Occlusion of lymph vessels is sometimes encountered during bouts of inflammation from various infectious agents, particularly those of bacterial origins. Isolating and identifying the filarial secretion(s) that reverses this swelling could eventually lead to a useful therapeutic agent.

As we learn more about the intimate lives of these interesting creatures, more uses for their secrets will become obvious—of that I am confident.

4 A PARASITE FOR ALL SEASONS

Toxoplasma gondii

When it comes to parasitism, even Carly Simon would agree that nobody does it better than *Toxoplasma gondii*. Nobody! The proof is in knowing the distribution of this obligate intracellular protozoan parasite. Members of nearly every species of mammal and numerous kinds of birds—in fact, all warm-blooded animals—have been found to harbor the infection. No other eukaryotic parasite can claim this degree of global distribution or breadth and depth of host range. What's more, in each infected animal, virtually every cell is susceptible to invasion by this highly promiscuous pathogen. While nearly all intracellular parasites (all viruses and most protozoans) interact with specific molecules on the surface of the host cell to ensure that they invade the correct cell type (e.g., brain cells, liver cells, kidney cells), *T. gondii* is a generalist and is able to infect nearly all varieties. What is more, it usually does it without doing much harm to its hosts. All it really wants in life is a nice, warm home and a long, sedentary existence. These facts alone make *Toxoplasma gondii* the *capo di tutti capi* of parasites!

Toxoplasma gondii is evolutionarily related to a large group of organisms called the Apicomplexa, which include the malarias and the diarrheal disease–causing protozoan *Cryptosporidium parvum*. While *T. gondii* is officially classified as a zoonosis (that is, it infects mainly wild animals), it is estimated that over 30 percent of humans (2.1 billion) are also infected. In some countries the rates in human populations are significantly higher. Take France, for example. Nearly 50 percent of the general population has experienced a primary infection with this parasite by the time they reach adulthood. Nearly all populations living at or above the Arctic Circle (e.g., Eskimos and Laplanders) are infected, and some villages approach the remarkable rate of 100 percent. In the United States the prevalence is lower, matching the global average of around 20 percent. We know all this because of data generated by extensive epidemiological surveys employing serological methods and sponsored by numerous health organizations, such as the Centers for Disease Control and Prevention and the World Health Organization, over the past ten years. These studies reveal the percentage of sera collected from around the world that contain specific antibodies against *T. gondii*. It is the best approach for determining the prevalence of toxoplasmosis in large populations, regardless of the species or geographic location.

What are the critical biological features of *T. gondii*'s life cycle that favor its spread over the entire planet, and what were the factors that enabled it to become the most popular kid on the block? Let's start with a detailed exposé of its modus operandi regarding how it gets from one host to another (fig. 4.1).

There are many ways an infectious agent can spread among a population of host animals: arthropod vectors (e.g., mosquitoes, black flies, kissing bugs, ticks, mites), sexual contact, aerosol spray (e.g., sneezing), fecally contaminated water and food, contaminated blood supply (i.e., transfusions), and congenital (mother to child). Amazingly, in the case of *T. gondii*, four transmission mechanisms are viable options for initiating infection in the warm-blooded side of the animal kingdom: water and food contaminated with cat feces, meat infected with the tissue cysts, congenital, and transfusion. The most common transmission cycle is carnivorism. Meat-eaters the world over help to distribute this protozoan from one group of scavenger/predators to another by the

Toxoplasma gondii

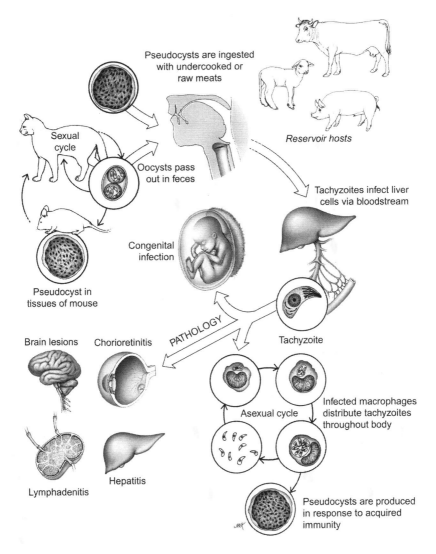

Figure 4.1 Life cycle of *Toxoplasma gondii*. Illustration by John Karapelou. (*Source: Parasitic Diseases*, 5th ed., (c) Apple Tree Productions, LLC, New York)

simple act of eating raw flesh. This implies that most animals will take a bite out of a carcass from time to time, depending on how hungry they are, even if their usual diet does not include meat. In nature there are very few strict herbivores—even hippopotami have been observed eating meat. That is why France, where steak tartare is a national dish, has such a high rate of infection, and those who eat raw meat exclusively (e.g., Eskimos) are almost universally infected.

While *T. gondii* can infect all mammals and birds, its definitive host is the domestic cat and its many wild feline relatives—the bobcat, tiger, lion, cheetah, panther, mountain lion. Recall that the definitive host is the one that harbors the sexual stages of the parasite. All other warm-blooded animals harbor the asexual stage and serve as the source of infection for cats and their relatives. Cats acquire toxoplasma when they prey on rodents and other wild animals that harbor the infective stage in their tissues. What's more, protective immunity induced by a primary infection with *T. gondii* is relatively short-lived in cats. Since all felines are primarily carnivorous, *T. gondii* stands a good chance of infecting or reinfecting these predators each time they dine al fresco. Domestic cats that remain housebound are far less likely to acquire the infection. But when the chef of the house prepares the evening meal and is in the habit of feeding scraps of meat to pets, they too can become infected. Mice in the house are another good source of infection for cats.

Most tissues of the intermediate host (i.e., any warm-blooded animal but cats), harbor the infectious stage of the parasite in a special entity called the tissue cyst. Inside each tissue cyst are hundreds of *Toxoplasma gondii* organisms called bradyzoites (the "brady" bunch). When the organisms are ingested, the thick wall of the tissue cyst is digested away, releasing the individual microscopic bradyzoites into the small intestine (see fig. 4.1). The parasites rapidly penetrate the cells lining the intestinal tract and begin their lives as intracellular organisms. In cats some of these newly acquired intracellular parasites develop into males and females. Then mating occurs, and the resulting stage emerges from the intestinal cell and transforms in the open space (lumen) of the small intestine into the environmentally resistant oocyst. When the oocysts pass down to the large intestine, they become incorporated into the fecal mass and are eventually eliminated during the act of defecation.

Outside the host the organisms inside the oocyst divide and develop into the infectious stage for all warm-blooded animals. To infect another host, however, the oocyst must be ingested. So, in addition to adopting a multipronged approach to its transmission cycle, toxoplasma has evolved two completely separate stages for infecting a host: oocysts and tissue cysts. These two general features of its life cycle elevate toxoplasma above all other parasites, making it the hands-down winner of the title "A Parasite for All Seasons."

Back inside the cat, bradyzoites that do not transform into sex cells have the option to develop into tachyzoites. Each tachyzoite can infect a new cell before replicating. In cell cultures *T. gondi* can produce from six to twenty new parasites (figs. 4.2, 4.3). In an infected animal, the number of new organisms produced by each round of parasite replication is not known, but it is suspected to be at least as high as the number produced in vitro in the laboratory. In the process, *T. gondii* kills the host cell on which it had feasted. When the host cell dies, tachyzoites are free to penetrate other cells and repeat the process. Tachyzoites cannot transform back into sex cells, however. Eventually, given enough cycles of replication, tachyzoites invariably infect cells that can travel away from the original site of the infection (e.g., dendritic cells and macrophages). That's when *T. gondii* becomes distributed throughout the body. As mentioned, no cell is immune to infection with *T. gondii*, with the possible exception of sperm and egg cells.

The ability of *T. gondii* to enter a cell is dependent on specialized organelles inside the parasite called micronemes and rhoptres, both of which are located in specific anatomic sites in the tachyzoite. Upon contact with a cell, a secretion termed the apical membrane antigen 1 (AMA1) is applied to the cell membrane and facilitates the interaction

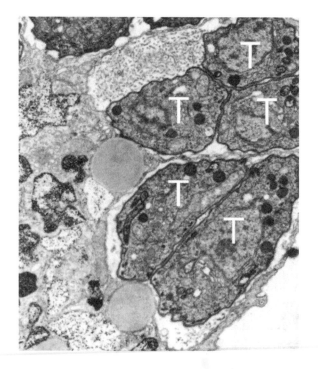

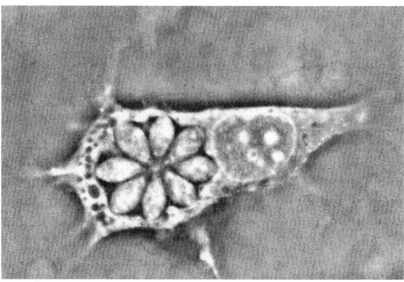

Figure 4.2 *(top) T. gondii* inside an infected cell. Electron micrograph.

Figure 4.3 *T. gondii* inside an infected cell

of the parasite's outer membrane with the host membrane, resulting in the formation of a tight junction between the two dissimilar layers. The junction aligns with host cell proteins involved in the movement of intracellular structures and with the anterior end of the tachyzoite. Other secretions from these glandlike structures empty their contents directly into the host cell. What happens next is pure poetry from the parasite's perspective. *T. gondii* glides into the host cell and is accepted, as if it was some long lost relative who just showed up on the doorstep.

So why can't we kill this organism with our remarkably well-developed immune system and eliminate it from our bodies, as we do to so many other would-be invaders? In fact, an infected host does not just sit there and take it. It responds with two arms of the innate immune system that collaborate to eventually slow down the rate of parasite replication. But regrettably, neither arm of the immune system is able to deliver the deathblow that knocks off the offending organisms (i.e., creates a sterile immunity). Instead, the host and parasite fight it out toe-to-toe for several weeks, eventually coming to a stalemate. The peace agreement includes benefits to both participants in the relationship: toxoplasma agrees not to replicate and thus ceases to do further harm, as long as the host agrees to harbor dormant (note I use the word *dormant)* parasites in tissue cysts scattered throughout the recesses of its body for the rest of its life. Skeletal muscle and brain cells seem to favor the development of tissue cysts, which seem to be able to live longer in these two tissues as well. Unfortunately, no relationship is ever truly 50-50. The fine print of this particular deal with the devil is as follows.

The immune system responds to the infection by ramping up the production of interleukin-12 (IL-12), a special, Paul Revere–like messenger peptide produced by certain immune cells and secreted into the bloodstream. This enables the IL-12 molecule to disseminate rapidly to all parts of the body, heralding, "The parasites are coming. The parasites are coming and they're infecting all our cells!" Interleukin-12 is also the poison apple that puts the parasite to sleep. As more and more

IL-12 is produced, the parasites become lethargic and susceptible to the inhibitory effects of a protein called gamma interferon (production of which is mediated by IL-12). All cells in the body are capable of making an interferon response to a broad spectrum of intracellular parasites. Together, these two products of the innate immune system are responsible for sequestering toxoplasma within all infected cells. However, as mentioned, this organism is not killed by this one-two punch. Instead, it responds with its own special survival tactic by modifying the infected cell membrane that surrounds it, eventually forming the outer wall of the tissue cyst. Try as we might, we cannot muster a stronger immune response that would wipe out the organisms inside the tissue cyst. That is because toxoplasma organisms go into suspended animation—a torporlike state—by slowing down their own DNA synthesis. Then they just hunker down for the long haul and do a parasite impression of Rip Van Winkle. This could last years; even longer in humans. Just how they manage to adjust their metabolism to the couch potato setting is not known, but we do know that if an infected host fails to maintain adequate levels of IL-12, these sinister sleeping beauties wake up and go back into hyperdrive. When they do, they destroy the tissue cyst wall and emerge out of their biological prisons as rapidly dividing tachyzoites to raise havoc with the host cells in their immediate vicinity.

Just such an occurrence was experienced on a massive scale during the advent of the AIDS epidemic in the 1980s. Individuals who became HIV-positive often died from encephalitis, a condition brought on by the reactivation of tissue cysts in their brain tissues. Coincidentally and highly regrettably, infection with the HIV virus down-regulates the production of IL-12. That is why the HIV virus is able to reproduce inside our T cells without hindrance from gamma interferon production. After numerous deaths occurred from toxoplasma-induced encephalitis, researchers developed a new strategy for treating HIV/AIDS patients, in which it was assumed that everyone diagnosed positive with the HIV virus was also infected with toxoplasma. At first all HIV/AIDS patients were treated for toxoplasma with bactrim, a combination drug that kills actively dividing *T. gondii* organisms by interfering with the parasite's DNA synthesis. Deaths from reactivated toxoplasma infection occurred only rarely thereafter. Today two other drugs, pyrimethamine and sulfadiazine, have replaced bactrim and are now the standard treatment.

Antibodies are soluble proteins that combine with offending compounds from a wide variety of sources, including microbes, and are also produced in response to infection with toxoplasma. How unfortunate, therefore, that they are unable to reach the organisms and kill them. Because the parasites live inside our cells antibody proteins are too big to cross the infected cell's outer membrane and get to where the action is. However, antibodies can prevent the spread of the organism from an infected mother to her developing fetus, but success is all about timing. If the antibodies are produced before the parasites get their opportunity to pass across the placenta, the developing fetus is spared. This typically is the case in France, where over 50 percent of young women have already encountered the infection. Antibodies stay elevated for years thereafter owing to the occasional breakdown and death of a tissue cyst. But if the woman has never been infected, becomes pregnant, and then acquires toxoplasma infection, it becomes a race against time as to whether her developing fetus will become infected. It all depends on how big the placenta is and how many toxoplasma tachyzoites there are inside infected cells (e.g., macrophages) that can cross the placenta. In the first trimester, weeks 1–12, the placenta is relatively small in mass and diameter, and so the chances are better than 90 percent that protective levels of antibodies against the organism will be achieved before an errant infected macrophage decides to wander across to the fetus from the mother-to-be. If the infection is acquired during the second trimester (weeks 13–28), the fetus and the placenta have dramatically increased in size. Thus there is a better than 50-50 chance that protective immunity will not develop before at least a few infected cells cross to the fetus, causing infection in the developing embryo. If the expectant mother becomes infected during her third trimester (weeks 29–40), then it is almost a sure bet (80–90 percent) that her fetus will become infected too.

Congenital toxoplasma infection, while rare in occurrence, can lead to serious consequences, with numerous organs becoming severely damaged or dysfunctional. Death of the fetus is typical if infection is acquired during the first trimester. If the fetus survives to term, the head of the newborn may be hydrocephalic. Before the discovery of human infection with *T. gondii*, as luck would have it, a hydrocephalic baby was diagnosed with this organism and became the first case of the infection reported in the medical literature. Because of the rarity of the condition, it was assumed that infection with *T. gondi* was also extremely rare. Only when we could actually survey large populations for the presence of antibodies against the infection did we learn that it was just the opposite of what we suspected. Today hydrocephalus due to *T. gondii* continues to be rare, while other pathological effects of congenital infection are more common.

For example, later on in the pregnancy, if the fetus becomes infected, say around week 30 or so, its immune system has already begun to form. This is important, since this developmental milestone can help to moderate the pathological consequences of the infection. Eye damage (chorioretinitis) and varying degrees of learning deficit are the expected outcome of infection acquired by the fetus during the late second and third trimesters. Regular testing immediately after a woman becomes pregnant allows the physician to monitor whether or not she has become infected. Since domestic cats are prone to infection, regardless of their past history of infection with toxoplasma, avoidance of any duties related to cleaning the kitty litter by the expectant mother is highly recommended. Freezing all meats before eating them kills the tissue cysts.

Regular checkups at the local veterinary clinic that include a stool examination are the best way to detect early acquisition of *T. gondii* by domestic cats.

Epidemics in nature among certain marine mammals have emerged in recent years, most notably among sea otters along the northern parts of the California coastline. The mortality rate due to infection with *T. gondii* in this group is high, although exact figures are lacking. The increase in sea otter infections is the direct result of degradation of coastal estuaries caused by human encroachment (farming, settlement). Feral cats abound in these damaged ecological settings, preying on displaced animals (mostly small rodents of various types) seeking refuge in nearby natural settings. Mud flats created by removal of protective vegetation make displaced animals easy targets for these hungry, agile, semiwild felines. When it rains heavily, the environmentally resistant oocysts are washed from these barren areas into the nearby marine environment and the otters pay the price, often with their lives. Furthermore, only sea otters that eat marine snails on a regular basis are at high risk of infection, while those animals that maintain a traditional diet of abalone are essentially toxoplasma free. Depleting the abalone resource forces otters to switch to other food items, including snails. This is additional evidence in support of the view that damaging marine coastal wetlands and harvesting essential sea otter food items combine to favor the transmission of *T. gondii* to local populations of marine mammals. As natural environments become more fragmented, other transmission cycles are created by the juxtaposition of wild felines with domestic cats. For instance, bobcats and mountain lions were shown to acquire their infections from coming into close contact with densely populated areas.

One unusual outbreak in humans in Atlanta, Georgia, was traced back to a riding stable. Rain had forced a group of dedicated equestrians inside to practice on that particular day. The hay used to feed the horses was stacked at one end of the stable and was home to thousands of mice. Three cats were kept around by the manager of the facility as a volunteer mobile rodent control system. But let's face it, how many mice can one cat eat each day? Not enough! The day's riding lessons were long and exhausting, one member recalled. Thirty-seven members eventually became infected; forty-nine other riders did not. It was shown later that of the three cats, two were sero-positive for *T. gondii*. The cats lived at the hay end of the stable, the same end where the infected riders rode. It was deduced later that probably the dust kicked up by the horses aerosolized the oocysts found in cat feces, and the unfortunate patrons had inhaled them. Eventually the organisms ended up in their small intestines. Fortunately, no one became seriously ill.

Plowshare Concept

What have we learned so far from studying the biology of *Toxoplasma gondii* that we might be able to expand on for our own benefit? How about the fact that this parasite can regulate its own DNA synthesis in response to host immune responses? Going into a state of dormancy suggests that we might be able to apply this strategy to certain types of difficult-to-treat cancers. The mechanism of cessation of DNA synthesis that *T. gondii* uses to avoid destruction by host defenses is reversible; witness the emergence of tachyzoites from the tissue cycts after IL-12 levels plummet. So if we could identify and replicate the mechanism in experimental settings, we may be able to achieve some degree of control over metastases. We could essentially turn the cancer on or off whenever we so desired. This suggests that this approach might also be advantageous in applications to diseases involving the overproliferation of cells in noncancerous situations as well (e.g., psoriasis, obesity, type 1 diabetes). Only further research will show whether these ideas have merit. Without new information and knowledge, we will be relegated to groping in the dark for answers to some of our most important questions.

Again we come to the crucial importance of funding for new scientific research on parasitism. What a shame if we choose to give up the good fight and leave these fruitful and exciting ideas unexplored, especially now when we are so close to revealing the answers to many of our other questions about how life works.

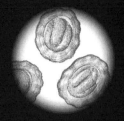

5 THE UNHOLY TRINITY

Ascaris lumbricoides, Trichuris trichiura,
and the Hookworms

Shortly after life arose, around 3.55 billion years ago, all of Earth's free-living organisms were soon face to face with pathogens. Most likely, these early parasites were viruses. As life continued on its inexorable journey toward complex multicellular creatures, these parasites, not wanting to be left behind, followed suit and evolved right alongside the other organisms. The well-worn expression "Nature abhors a vacuum" rings particularly true when we try to imagine how many species must have lived over the billions of years that life has existed on this planet. Even more mind-boggling, at least to a parasitologist, is trying to envision the sheer number of opportunities life must have presented to all varieties of parasites, inviting them to come on in and settle down into a comfy new home. Today there is a robust array of these interlopers, ranging from the ultrasmall (viruses, bacteria, fungi, protozoa) to the ultralarge (worms), whose ecological role in nature is well defined: to limit the size of host populations by reproducing like crazy, at their host's expense.

But life is not all one-sided. It is also about balance and give-and-take. Parasite infections, no matter how lethal they may be for any given host species, tend to spare those rare, preadapted individuals, which, by virtue of random mutations, are able to tolerate their parasites' pathological consequences. In other words, if an epidemic is severe enough, the majority of the host population might well succumb to it, leaving behind only those few resistant organisms to repopulate. When the same epidemic recurs after some years have passed, practically all the members of that newly reconstituted population will survive. Such was the case in Europe, for example, during the early plague years.

But random mutations also occur in the parasite population, occasionally resulting in an even more lethal variety. If this should occur, the balance once again might swing back in favor of the parasite, not the host. These oscillations between deadly epidemics and resistant host populations have been likened to an ever-escalating arms race between two superpowers. Unlike modern warfare, however, in nature there are usually no clear winners in the long term, as the physical environment and the chemistry of the DNA molecule continually reset the stage for the next round of face-offs between various life forms. In other situations, the lethality of a given pathogen may actually decrease the more it passes among large groups of hosts. The result is a diminished rate of mortality, and in extreme cases the offending microbe may actually transition into a friendlier symbiote that then goes on to live in harmony with its host. That is probably the mechanism by which we have come to accommodate so many bacterial and viral entities. For example, the current literature lists some 1,500 different species of bacteria as *normal* inhabitants of our small and large intestinal tracts. Without them, the simple act of defecating might well be redefined as a situation requiring medical intervention.

There are many other examples of give-and-take between us and our pathogens that have occurred numerous times throughout our natural history, and I have mentioned some of them in earlier chapters. Another example that comes to mind as strongly supporting the argument above occurred after the Spanish conquistadores invaded the Americas in the early 1500s. When the Spanish empire finally abandoned its imperialistic folly in the late 1600s, nearly forty-five million natives had perished, and not from swords and bullets either, but, incredible as it seems, from common infections their immune systems had never before encountered (e.g., influenza, chicken pox, measles, puerperal fever), brought to them courtesy of gold-hungry marauders from the Old World. Within a hundred years after the collapse of the Spanish empire in the New World, the Americas had rebounded, in fact far exceeding their original population density. That is because the Spanish left behind their genes. The resulting new generations of hybrid individuals now could resist the introduction of infectious agents common to the Old World. This same general scenario has played out again and again in different parts of the globe, orchestrated by outbreaks of plague, small pox, malaria, cholera, and the infamous 1918 influenza epidemic that alone laid waste to an estimated 2 percent of the world's human population.

It is true that these organisms have often brought us to our knees and, in a few instances, have had us teetering on the brink of extinction. But today, although many of these organisms are still with us (except for small pox) and remain an integral part of our natural heritage, there are important differences between now and then that are worth pointing out. The advent of sanitation, a passel of effective vaccines, and a cornucopia of chemotherapeutic agents, all brought to the fore just within the past 120 years, have altered the very process of natural selection in favor of those populations lucky enough to be able to afford these technologies. In each successful instance, that culture has flourished and prospered well beyond any other time in its recorded history.

But not everyone has had the advantage of living a life essentially free of infectious diseases. In fact, over half of the world still lives much as we all did before the modern age of public health and science-based medicine came to our rescue. In that less fortunate, more naturalistic world, the forces of natural selection are still hard at work, unabated and raw-edged. Tuberculosis, HIV/AIDS, and malaria remain as major detractors of human health wherever economics dictates that vast numbers of humans cannot afford treatment or to institute long-term preventive strategies. These three plagues on our house have not gone away just because we can now cheaply produce 70-inch plasma-screen televisions or fly anywhere in the world in just under eighteen hours. But while a handful of high-profile diseases have finally risen to the top of our priorities list of things to address, there still lurk about in that darker world other, less heralded, less aggressive pathogens that have hung on and become part of the everyday world for billions of people. These are the most underappreciated of our parasites that have for centuries kept a low profile, cruising just under the radar screen of the world's health agencies. That is, until now.

Taken as a whole, these other persistent infectious agents have for centuries been directly responsible for holding back the social and economic advancement of countless societies throughout the less-developed world. All are worm infections, and three of them are roundworms, referred to by some as the "unholy trinity." Others call them part of the "great neglected diseases." They are the giant intestinal worm *Ascaris lumbricoides*, the whipworm *Trichuris trichiura*, and the two hookworms *Ancylostoma duodenale* and *Necator americanus* (see chapter 2). The other members of this gang o'worms include the schistosomes and the filariae (see chapter 4), and among this gang only the filariae are transmitted by insects.

No matter what you call them, though, these worms remain indifferent to us and resolute in their singular mission. They will not easily go away. That is because, for the majority of them, the transmission route for getting from one host to another is dead simple. Their transmission cycles are solely dependent on contamination of our food and water with soil that contains their embryonated eggs or larvae. Because the soil is the vehicle for getting them to us, these worms are referred to as "geohelminths"—literally, "ground worms." In the past medical anthropologists have been guilty of characterizing such human habitation that has high infection rates with judgmental descriptors like "filth and squalor." Today, on the other hand, sociologists and public health professionals have come to realize that these worm infections are linked inexorably to those cultures suffering from poverty and high rates of illiteracy. Change these ecological social conditions to those found in the developed world and none of these worms would be tolerated in their environment for even one moment.

Their presence is generally viewed as a symptom of regression in our efforts to live healthy, productive lives. Our inability to properly dispose of feces and urine continues to haunt us and give us grief. However, one cannot blame people living in abject poverty for their well-intentioned behaviors that also just happen to lead to a reinforcement of certain parasite life cycles. That is because feces are a commodity highly valued by over half the world's farmers. They are a rich source of nutrients for crops like rice, barley, and leafy green vegetables. Commercial fertilizers are out of the question, being far too expensive to even contemplate using. So, for centuries, farmers in Africa, Central and South America, India, China, and Southeast Asia have spread the by-products of their own metabolism onto the growing fields to regenerate yearly crops of rice, corn, barley, and a wide variety of vegetables. Unfortunately the eggs of certain worms are designed to survive for long periods of time in soil, making that the ideal medium for contaminating harvested crops. The rest of the story is easy to piece together. Some two billion people around the world harbor combinations of these three intestinal worms, making them, indeed, the unholy trinity.

The Giant Intestinal Worm: *Ascaris lumbricoides*

Imagine a smooth, pink-tinted worm about the size of a pencil (fig. 5.1). Now imagine that this same worm just crawled out of your left nostril! Sounds like your worst nightmare, doesn't it? A truly disgusting, surreal experience. Yet that is actually what happened to the six-year-old female patient from Guatemala. Both she and her mother were admitted to

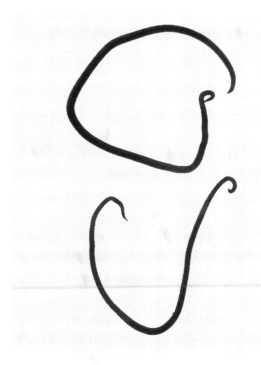

Figure 5.1 *Ascaris lumbricoides* adults. The male is small-er than the female.

Columbia/Presbyterian Medical Center in New York with HIV/AIDS, tuberculosis, and malaria. They were admitted based on the single di-agnosis of active TB, made at a neighborhood clinic. No further clinical information for either person was available to the house staff. Unbe-knownst to anyone, the daughter was also supporting many adult *Asca-ris lumbricoides* that were all stuffed into the anterior part of her small intestine. The day after she entered Babies Hospital, her malaria infec-tion flared up and she developed a high fever. That was the last thing the clinician wanted to have to deal with, given the complexity of the girl's condition, and without any information yet from the laboratory to help guide the approach to treatment. Luckily there were other caregivers around to help out. One of them was a third-year medical student fresh out of the course I taught in parasitic diseases. The student told me what happened next.

The attending resident in charge of the case (who, as it turned out, had graduated from another medical school at which parasites were hardly mentioned in any of the courses) essentially freaked out at the sight of the small child calmly sitting up in her hospital bed with a slowly writhing pink worm hanging halfway out of her nose. No one, nurses included, would go near the small girl. The medical student, now on her second week of clinical rotation in pediatrics, quickly sized up what was happening and took control of the situation. She first put on a pair of latex examining gloves and proceeded to remove the worm, all the time trying to comfort the patient, telling her that everything was going to be all right. "What the hell *is* that thing?" the resident screamed. The student, without stopping what she was doing, replied matter-of-factly that she thought it was an adult ascaris, but that she would personally deliver it to Miguel Guelpi, the parasitology diagnostic technician in the hospital's Microbiology Diagnostic Laboratory, for confirmation.

Meanwhile the student offered that the attending physician might want to give the child some aspirin to alleviate her fever until its origin could be determined, since, as she recalled, high fever was the most likely reason for the worm to start migrating out of the patient. She added: "Dr. Despommier told us in class that ascaris gets really *pissed off* and tries to leave the body when the patient's temperature goes up. The child might be harboring more worms." That same day, the laboratory tests confirmed the student's suspicion and antimalarial therapy was started, along with a course of mebendazole, the drug of choice for ascaris infection. I learned much later that, tragically, the child had expired. I never found out what happened to the mother.

At any one time there are some two billion people infected with *A. lumbricoides*. Multiple infections are common in places where fecal contamination of the water and food supply is a given, and insect vectors like mosquitoes can bite us whenever they feel the urge. Imagine, then, how many people living day-to-day under these difficult conditions also experience fever induced by a plethora of infectious diseases. Aberrant migrating adult ascaris must be very common indeed. What is even more sobering to realize is that when ascaris decides to relocate, any number of other possibilities could also occur, many of which have proven fatal. Worms have migrated into the liver, destroying the integrity of that essential organ. Adult worms have blocked the duct leading to the pancreas, rendering that organ inoperative. They have perforated the small intestine and induced peritonitis by inadvertently introducing bacteria from the gut track. Rarely (horribly), large numbers of migrating adults have been the cause of death by suffocation, migrating en masse up into the pharynx.

Several years ago I had the pleasure of visiting Bangladesh, and while there I was introduced to the faculty in the Department of Parasitology at the Bangladesh College of Physicians and Surgeons. We had a very nice exchange of ideas related to the teaching of parasitic diseases. The professors there said it was quite easy for them to teach by example since there were so many patients to choose from who showed up at the clinics each day harboring a wide variety of parasites. They then showed me a very large glass water jug, perhaps 50 gallons in capacity, filled to the brim with formalin-preserved adult ascaris collected from a single rural village during a one-day treatment/collection sortie by a group of third-year medical students.

The life cycle begins when we ingest the embryonated egg of ascaris (fig. 5.2). Once swallowed, it passes through the stomach and enters the lumen of the small intestine. The egg will not hatch until it has received specific environmental cues from the host. These include being exposed to rapid changes in the pH (acidic, then basic). Carbonated beverages, for example, are considered quite acidic, while milk and yogurt are at the opposite end of the pH spectrum, basic. These two conditions are

Ascaris lumbricoides

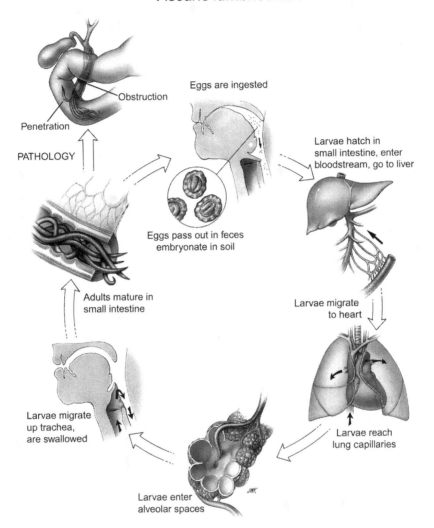

Obstruction

Penetration

PATHOLOGY

Eggs are ingested

Larvae hatch in
small intestine, enter
bloodstream, go to liver

Eggs pass out in feces
embryonate in soil

Adults mature in
small intestine

Larvae migrate
to heart

Larvae migrate
up trachea,
are swallowed

Larvae reach
lung capillaries

Larvae enter
alveolar spaces

Figure 5.2 Life cycle of *A. lumbricoides*. Illustration by John Karapelou. (*Source: Parasitic Diseases*, 5th ed., (c) Apple Tree Productions, LLC, New York)

sequentially maintained in our gut tract. This indicates to the worm that it has moved on from the stomach (which is acidic) to the small intestine (which is basic). Many kinds of worms have been selected for life inside us that are dependent on this common theme. Hatching cannot proceed unless these two pH changes are first experienced.

There are other amazing ways in which the worm has evolved to move seamlessly through the digestive tract. For instance, it so happens that an essential worm enzyme inside the egg is needed for hatching and cannot be activated unless one of the outer lipid (fatty) layers of the eggshell is first dissolved. This takes place in the upper small intestine, when the egg encounters bile. Then the eggshell becomes permeable to other chemical signals from the host, which activates the hatching enzyme that, in turn, dissolves the cement holding the lid of the eggshell in place. When the lid lifts off, the infective larva crawls out. Now it knows exactly where it is.

In nature, these processes between host and parasite are all in sync. Thus, the act of a parasite's egg hatching in the lumen of our small intestine appears to be a simple process, but it turns out to be quite involved, once all the facts regarding what it takes to get it to hatch are known. I find much beauty in nature at this level, and that is why I have remained passionate about understanding the finer details of how our parasites live within us. The deeper I look, the more beauty I find.

The microscopic larva of *A. lumbricoides* is now free to penetrate the wall of the small intestine and enter our bloodstream. At this point in its life, its diameter is the same size as that of our red blood cells, so passing through the capillaries is not a problem. Once in the bloodstream, the worm begins its travels throughout our body on a weird and tortuous journey and amazingly ends up right back where it began: our small intestine. Its first stop is the liver. It seems as though the larva of ascaris has developed a definite fondness for *foie gras de l' homme*. It feeds voraciously on liver cells for several days until it achieves its next growth stage. Now its diameter is larger than the capillaries of our circulatory system. This will prove to be crucial when it reaches those small vessels.

From the liver the worm reenters the circulation and eventually reaches the capillaries of the lung. But because the immature ascaris worm is now too big to pass through, it gets stuck in a capillary adjacent to the small air sacs that enable us to exchange oxygen and carbon dioxide (the alveolus). It actually feels the pressure of the capillary cells on its outer surface, and this stimulates the worm to penetrate out into the air sac. It crawls up our respiratory tree until it reaches the epiglottis and crawls some more until it is now once again in the esophagus, right where it started out as a larva inside its eggshell several days ago. When the host swallows, the larva is carried to the small intestine, where it stays put. Within the next six weeks, it matures to the fully grown adult worm (fig. 5.1) and remains there for up to five years.

To grow, it needs a source of nutrition. In this case it's a no-brainer. Ascaris adults eat what we eat. In fact, they get first crack at all our proteins because they secrete a factor that inhibits our own protein digestion. In some extreme cases, this can result in protein malnutrition for the infected individual. This is all too common among children with heavy worm burdens that may number in the hundreds.

We can thank the Koino brothers in Japan, who, way back in 1922, risked their lives to show what happened to embryonated ascaris eggs after they were swallowed. One brother actually ingested viable eggs of a closely related species, *Ascaris suum*, a very common infection of pigs, and easy to come by at any slaughterhouse. After maintaining the eggs stripped from an adult female ascaris for one week in a culture of finely divided charcoal, he counted them out under a microscope, transferred them to a test tube along with a small quantity of water, and swallowed the mixture. Six to nine days later, his brother analyzed his sputum and identified many migrating larvae of *A. suum* in it. None of the larvae that were not coughed up but were swallowed produced adult worms, since this species is better adapted to pigs than to humans. They then repeated the experiment with *A. lumbricoides* eggs in the same brother who had swallowed *A. suum* eggs. That infection also demonstrated that the lárvae migrate to the lungs before completing the cycle in the small intestine. This infection produced lots of adult worms. Fortunately there was a good drug available to treat the infected Koino sibling.

It should be now obvious to the reader that the field of parasitology attracts unusual people to it! Many of them conducted similar experiments on themselves with other parasites, but not always with the same benign results. In one infamous case an investigator infected himself with *Schistosoma haematobium* while working in Egypt so that he could have a ready supply of eggs to work on in his laboratory when he returned home. The infection proved fatal. Many years later it was learned through further experimentation in other laboratories that certain species of rats could also serve as competent hosts, thus obviating the need for self-inflicted parasitism with this potentially deadly worm.

Back to the ascaris, now living and feeding happily in the small intestine. Once the adult worms have matured (ninety days after being swallowed as an egg), mating ensues. About two weeks later the females begin to shed eggs. Each female ascaris produces about 200,000 eggs per day, and she can do this continuously for up to five years. Imagine how much nutrition in terms of total calories each worm must ingest from our diet each day to make that many eggs! Now multiply that number by the number of female worms infecting the person, then by 365, and that number by two billion, and you get some idea as to the global scope of the problem. It is entirely possible that over 25 percent of what the less developed world eats each day goes to feeding intestinal parasites.

Control or eradication of ascaris infections depends on containing the spread of human feces (i.e., good sanitary practices) because this worm only infects humans. There are no reservoir hosts. Despite this fact, ascaris continues be one of the most common infections in the world. The fact that its egg can survive in the soil for long periods of time enables it to remain a public health problem, despite our best efforts to limit its spread. Of course, mebendazole, the drug of choice for treating the infection, is readily available throughout the world, but cost is still often prohibitive where it is needed most. Various international health agencies (e.g., the World Health Organization) and many nonprofit organizations have sponsored community-based intestinal worm control programs in which a cocktail of antiparasitic drugs is administered to children in schools throughout a prescribed period of time, usually once a year for several years. These have been variably successful, in part owing to high rates of reacquisition of the infections. There are no vaccines for any worm infection, including ascaris. Although there are currently several effective, affordable (for most) drugs to treat ascaris infections, it is inevitable that this worm will one day become resistant to them. The only successful long-term strategy for controlling this infection is a good sanitation system.

The Whipworm: *Trichuris trichiura*

The second member of the unholy trinity is *Trichuris trichiura*—more commonly known as whipworm. Its prevalence is a bit less than that of *Ascaris lumbricoides*, and it does not have the same clinical drama that a migrating ascaris does—although, as I will relate below, I do know of one rather gruesome clinical appearance of whipworm. Do not underestimate the whipworm's ability to inflict suffering and pain; trichuris is a force to be reckoned with.

An estimated 1.5 billion people are infected with "tt," as the clinical parasitology lab techs refer to it. During my time at Columbia Medical Center, we saw a lot of tt among our upper Manhattan patient population, who came largely from the Caribbean. Because of the usual clustering of parasite infections, as I mentioned above, it was an unusual day when we saw ascaris eggs under the scope in a stool sample from a patient and did not also see eggs of tt on the same slide. In those days (the 1960s), nearly 25 percent of people from the Dominican Republic alone were positive for whipworm. Most were children between the ages of five and fifteen years. Its global distribution also closely follows that of the other two parasites in Club Geohelminth.

Trichuris trichiura is called the whipworm for a very good reason: it is shaped like a whip. The head end is the narrow whip part, and the tail is the handle. The entire adult worm measures around 40 millimeters in length. But unlike ascaris and hookworms, which live in the nutrient-rich small intestine, trichuris resides further downtown, in the large intestine, with its head end firmly embedded into the tissue (fig. 5.3). Trichuris adult females (fig. 5.4) produce eggs that need to sit around in soil for at least a week at room temperature before they embryonate and become infectious for the next host, just like ascaris. The male of trichuris is smaller (fig. 5.5), with a curved tail, following the same pattern as ascaris adult worms. Its life cycle is shown in figure 5.6.

The diseases whipworm causes depend on the severity of the infection. A few worms are well-tolerated, even in very young children. This is in stark contrast to ascaris, where even one fully grown parasite that migrates into a vulnerable anatomic location can really cause trouble. If more than twenty *T. trichiura* are present, a toddler might experience mild diarrhea. With even more worms (hundreds), the diarrhea may evolve into dysentery (bloody diarrhea), accompanied by anemia. Trichuris infection, no matter how heavy the worm burden, rarely proves fatal. That is its mantra: "injure, but don't kill." It needs the host to remain alive or the parasite will die too. The anemia prevalent in heavy infections is not directly attributable to the dysentery, however. It is linked to the presence of the adult worms and their effects on the bone marrow's ability to produce red blood cells. As of yet, no specific compound from the worm has been implicated as the cause of the anemia. Trichuris adults can live for up to two years and can be acquired again and again by small children between the ages of two and ten until

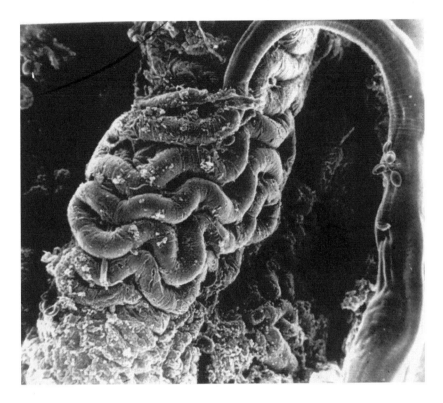

Figure 5.3 *Trichuris trichiura* embedded in the large intestine. Scanning electron micrograph.

their immune systems mature. That is why it is rare to find heavy infections of this parasite in adults living in the same environment.

Speaking of the clinical effects of heavy infection with trichuris, the following incident was shared with me by a renowned pediatrician and tropical medicine expert. After graduating from medical school, he served four years in the United States Navy as a pediatrician, caring for children of service members. While in transit from Japan to another port of call in East Asia, the ship picked up some Japanese civilians. Among them were a mother and child from one of the outer islands. He observed that the two-year-old daughter was irritable and crying

Figure 5.4 Adult female *T. trichiura*

Figure 5.5 Adult male *T. trichiura*

as they boarded; nothing unusual here. But as soon as they got under way, the mother came to sickbay with her child in tow and related through a translator that she often "saw worms" on the infant's bottom. Again, nothing odd about that, considering that pinworm (*Enterobius vermicularis*), the most common worm infection known to humankind, is ever-present in that age-group. "No," she insisted, "big worms on her

Trichuris trichiura

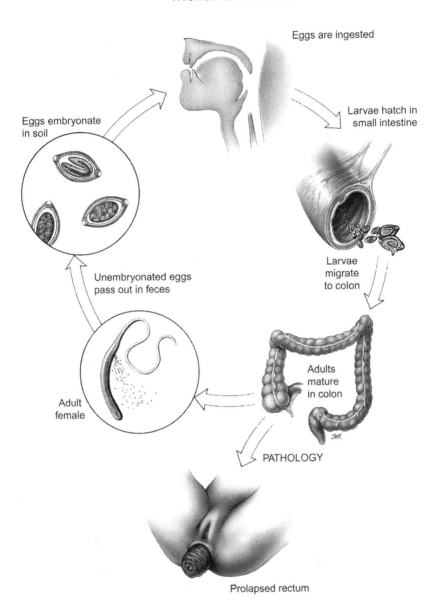

Eggs are ingested

Larvae hatch in
small intestine

Eggs embryonate
in soil

Larvae
migrate
to colon

Unembryonated eggs
pass out in feces

Adults
mature
in colon

Adult
female

PATHOLOGY

Prolapsed rectum

Figure 5.6 Life cycle of *T. trichiura*. Illustration by John Karapelou.(*Source: Parasitic Diseases*, 5th ed., (c) Apple Tree Productions, LLC, New York)

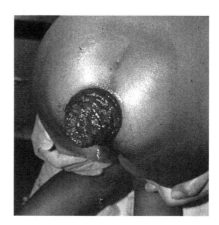

Figure 5.7 Prolapsed rectum caused by
T. trichiura

bottom!" When the doctor removed the diaper, lo and behold, there was
the child's rectum, everted and inflamed (fig. 5.7). What's more, it *was*
covered with big worms! The young clinician consoled the mother as
he slipped on a pair of examining gloves and pushed the rectum back
in place. He knew that severe infection with trichuris could result in a
prolapse of the rectum, having seen one example of that in a rural pe-
diatric clinic as a fourth-year student on a medical elective in Colombia.
He assured the mother that treatment of the infection would solve the
problem. She smiled and left the room confident that her child would
be all right. After he closed the door, he did what every other U.S. naval
officer likely would have done in that situation: he heaved up his lunch
all over the examining table and couldn't even think about food for the
next two days.

The life cycle of trichuris is very simple (fig. 5.6). Once an embryonated egg of tt is eaten, the parasite receives its set of environmental
cues from the host as it passes down the digestive tract. It hatches in
the small intestine and then embeds itself in the tissue and proceeds
to grow into an immature adult. It then crawls out of the tissue and is
carried down to the large intestine where it re-embeds into the tissue of
the intestinal wall, and there it remains for the rest of its life. Compared

with ascaris, a worm that swims freely in the small intestine and can go anywhere it chooses, trichuris is a real stay-at-home, content to enjoy the comforts of a domicile that it creates for itself. Somewhere between the small and large intestine, mating occurs. When the adult female worms mature, they begin to pass eggs. Only the head end of the worm is embedded in the tissue (fig. 5.3), leaving the tail end dangling freely in the lumen of the large intestine. Eggs exit from a pore located at the junction of the head and the tail. The entire process, from the start of the infection to the first day of egg production, takes about ninety days.

Trichuris trichiura is part of a large family of related roundworms and is a distant cousin of *Trichinella spiralis* (see chapter 1). There are species of trichuris that infect dogs (*T. vulpis*), sheep (*T. ovus*), pigs (*T. suis*), and rodents (*T. muris*). They all behave similarly and have nearly identical life cycles as well. This makes trichuris easier to study if we want to know more about its biology than just the kinds of diseases it can cause. So what do trichuris adult worms eat? How do they feed? Well, it's somewhat embarrassing to admit, but parasitologists are still trying to answer these basic questions. The adult worm is similar in its anatomy to that of trichinella. Recall that the adult of that other worm lives embedded in a row of cells in the small intestine, and we don't have a clue as to its nutritional needs or its mechanism of feeding either. It fact, it is fairly safe to say that we have very little information regarding these essential details for the whole family, of which trichuris and trichinella are but a small part.

This suggests that these kinds of infections are indeed among the great neglected diseases. Research dollars are limited, and it is therefore quite understandable that funding priorities must target those infections that have the greatest global impact on morbidity and mortality. That is why research on most worm infections (the schistosomes excluded) continues to get the short end of the funding stick. Yet a study conducted by the London School of Economics concluded that worldwide, more days of work are lost to illnesses caused by the gang o'worms than by HIV/AIDS, malaria, and tuberculosis combined. The maddening thing to realize is that the simple application of low-maintenance sanitation technologies to most high-transmission zones throughout the tropics for these worms would solve the problem once and for all. Granted, we would still need to address vector-borne diseases, including filarasis and malaria, but there are reliable findings that show that these soil-transmitted worm infections make us more vulnerable to the pathological effects of malaria, TB, and HIV/AIDS. Solve the sanitation problem first, and the rest of those Most Wanted outlaw parasites and infectious diseases become much more manageable.

All of what we now know about the clinical effects of trichuris was exceptionally difficult to ascertain since most patients with trichuris that were studied had other intestinal parasites as well. Most were host to at least one other member of the unholy trinity, and they usually had other entities, too, like giardia and any number of other intestinal protozoan infections that also caused things like diarrhea and weight loss, making it impossible to sort out what only trichuris was doing to them. Eventually, though, enough patients harboring only tt were identified and studied, giving physicians a clearer picture of what this worm's effects are on children's well-being. We have finally confirmed that long-term infection in children can lead to malnutrition, stunting, and loss of cognitive skills. Coupled with what we already know about the diseases caused by hookworms and ascaris, these sinister consequences reaffirm the idea that it's crucial to try to address the problem of worms that affect the lives of over half the world's population.

Plowshare Concepts

Some of the most interesting findings with regard to intestinal parasitic worms are their connection with the hygiene hypothesis. This recently conceived notion is based on the observation that, worldwide, people living within high-transmission areas for these worm infections have dramatically fewer allergies to common things such as ragweed pollen, grasses, and foods such as shellfish and peanuts, and have greatly reduced rates of asthma and other even more serious autoimmune diseases, for example, Crohn's disease. In addition, many people suffering from these ailments apparently are less ill or not ill at all after purposely being given an infection with a small number of hookworms. These data suggests the following:

> Human evolution with hookworms has reached a stage where humans are relatively asymptomatic when harbouring low-intensity infections, assuming reasonable nutritional status of the host. Evidence is gathering that the hookworm manipulates the human immune system such that the infection is tolerated with minimal pathology to either the worm or the host.
>
> It seems likely that during our coevolution, the immune system has adjusted to compensate for the continual suppression by hookworm infection. Thus, in the absence of these parasites, our immune responses have become "hyperactive," resulting in an increase in the prevalence of immune dysregulatory illnesses in the developed world. Future studies will show whether we can use hookworms, or preferably molecules derived from them, to correct this imbalance. Indeed, if vaccines and other control measures aimed at reducing the prevalence of hookworm (and other neglected tropical diseases) are implemented en masse, the resulting effect on the prevalence of autoimmunity and allergy in these countries is of potential concern. (H. J. McSorley and A. Loukas, "The Immunology of Human Hookworm Infections," *Parasite Immunology* 32 (2010): 549–59)

Trichuris suis, the pig version of whipworm, has also been tried with some success in treating patients suffering from Crohn's disease. This approach was predicated on the unusual finding that almost no one suffers from that disease anywhere throughout sub-Saharan Africa, whereas their parasite-free relatives, now living for several generations in the United States, have rates of this disorder similar for those people who have no genealogical connection to Africa. Thus a hypothesis arose suggesting that the true targets of that disabling immune disorder were our old nemesis, the intestinal worms. Proof of concept was difficult to obtain, as most institutional review boards at every research-oriented medical school remained understandably skeptical regarding the risks and benefits of such an approach and predictably denied numerous researchers' requests to be the first to conduct such a study. Finally, enough data accumulated from the epidemiology literature to warrant a full-scale test of the hypothesis, and like so many other far-out ideas, this one struck pay dirt. The following is an abstract of a peer-reviewed and published set of studies produced by two gastroenterologists who championed this approach. They based their conclusions not only on epidemiological evidence but also on solid laboratory experimental results using mice and their worm infections to modify diseases such as type 1 diabetes, colitis, and asthma.

There is an epidemic of immune-mediated disease in highly-developed industrialized countries. Such diseases, like inflammatory bowel disease, multiple sclerosis and asthma increase in prevalence as populations adopt modern hygienic practices. These practices prevent exposure to parasitic worms (helminths). Epidemiologic studies suggest that people who carry helminths have less immune-mediated disease. Mice colonized with helminths are protected from disease in models of colitis, encephalitis, Type 1 diabetes and asthma. Clinical trials show that exposure to helminths reduces disease activity in patients with ulcerative colitis or Crohn's disease. This chapter reviews some of the work showing that colonization with helminths alters immune responses, against dysregulated inflammation. These helminth-host immune interactions have potentially important implications for the treatment of immune-mediated diseases. (D. E. El-liot and J. V. Weinstock, "Helminthic Therapy: Using Worms to Treat Immune-Mediated Disease," *Advances in Experimental Medicine and Biology* 666 (2009): 157–66)

If we could identify the worm components that the immune system targets, a treatment strategy could be developed that does not require the use of live worms. It should be pointed out that eventually all of those receiving therapeutic doses of worms rejected their infections and were therefore immune to reinfection, thereby terminating the therapy. This unfortunately allowed their original allergic conditions to relapse. Using a worm product (i.e., a nonliving component) to do the same job would allow for a longer treatment period, even if a mild form of immunosuppression was also needed to keep the patient from neutralizing the parasite molecule.

Hookworms have been tried in hopes of modifying other immune disorders, such as celiac disease, an allergy that many people develop to wheat gluten protein, and type 1 diabetes. In this case, however, the infection had no measurable effect on the status of the allergy when the infected patients were given a challenge of wheat glutin. The jury is still out on the efficaciousness of hookworm infection in alleviating the ravages of type 1 diabetes.

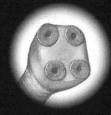

6 THE LONG AND THE SHORT OF IT

Tapeworms—*Taenia saginata, Taenia solium, Diphyllobothrium latum, Echinococcus granulosus,* and *Echinococcus multilocularis*

Tapeworms get their common name from their off-putting resemblance to a white cloth tape measure. But while working in the parasitology diagnostic laboratory during my formative years as a budding microbiologist, I came to know them from an entirely different cultural context: food look-alikes! The adult worms actually present as yellowish-white, long, flat ribbons. When one of the more robust species is piled onto a white shallow dish for examination, it resembles in every way a plate of pasta without the marinara sauce. The only difference comes if the worm is freshly obtained, and thus still alive. Then the dish of pasta slowly undulates and writhes.

Owners of pet cats and dogs are more likely to have had an occasional encounter with a tapeworm and so are sometimes more at ease with these creatures than are those who grew up in the absence of domestic animals. But for many (myself excluded, of course), tapeworms remain among the most reviled and misunderstood members in nature's menagerie of infectious agents. For some they conjure up phantasmagoric visions of alien biology lurking in our intestines, sucking up all our nutrients. But as I will describe later in this chapter, some of the most dangerous, even life-threatening members of this lowly life form are barely visible to the naked eye.

They are the frequent subject of pejorative remarks directed at overweight people and beanpoles alike: "That guy could use a couple of tapeworms"; "He eats like he has a tapeworm." Even Mark Twain had something to say about tapeworms: "Only presidents, editors, and people with tapeworms have the right to use the editorial 'we.' " You'll recall that that's the second quote about parasites from Samuel Clemens I have used in this book. Maybe that great thinker and humorist might have become a parasitologist had he not taken up his calling as the public voice and conscience for the average American. Come to think of it, I also remember him saying: "There is no distinctly American criminal class—except Congress." Twain would have made a helluva parasitologist.

But it was a few of the larger members of this group (that august electorate aside) that catalyzed the push in the 1700s toward a scientific understanding as to what constitutes a parasite. This was due to the tapeworm's common occurrence throughout most European communities. Some of them exceeded 20 feet in length as fully grown worms! Hard to miss an object that big. Around that time, curious minds began to ponder how these slimy critters carried out their lives. It was not until the 1800s, however, that progress was made in connecting a seemingly disparate set of dots to show that all of them required several different kinds of host to complete their life cycles, and that to acquire one, something had to be eaten.

Today we know quite a bit more about how these parasites manage to live such long and rather sedentary lives (often exceeding twenty years) at our expense. All vertebrates (fish, reptiles, amphibians, birds, mammals) harbor their own species of tapeworms, which live as adult worms in the lumen of their host's intestinal tracts. Most species of tapeworm are restricted to only one host species as an adult worm infection, but some zoonotic forms (i.e., ones that routinely infect animals other than humans) can occasionally cross that barrier and infect us as well. As one might guess, even without the advantage of getting the inside scoop from Jonah himself, whales hold the world's record for providing a home to the longest tapeworm—a specimen exceeding 120 feet in length. Just how this particular species of tapeworm makes its way into the intestines of an oceangoing mammal is still a matter of conjecture. So, as interesting as many of the tapeworm stories are for those species that infect wildlife, I will restrict the discussion that follows to the ones that routinely infect humans.

Tapeworms occupy a branch of the tree of life under the following headings: *Invertebrate* (no backbone)–*Platyhelminthes* (all flatworms)–*Cestoidea* (kinship groups). None of them have direct life cycles. That is, we cannot acquire an adult tapeworm simply by eating the tapeworm's eggs. Instead, we have to ingest an intermediate host, or a portion of one that harbors the infectious immature parasite. Each species of tapeworm employs a different intermediate host to get from one definitive host to the next. On the surface, this sounds too complicated a strategy to be an effective means of transmission. But judging by the global distribution of tapeworm infections, and the number of different host species that harbor them, parasitologists now consider the cestodes to be among the most successful of the worm infections. Many of the tapeworms infecting humans are acquired by eating an immature stage living inside another mammal. The fish tapeworm, *Diphyllobothrium latum*, is an exception to this rule. When I discuss its intricate life, the reader will see right off what is meant when I say it has a complex life cycle.

Let me first summarize the commonly shared features of tapeworm biology before exploring the lives of specific species. Each adult tapeworm consists of several parts, linked together in a chain of segments referred to as *proglottids*. The segments are produced by the neck region of the worm that lies just downstream from its head, called the *scolex*. The worm uses its scolex to attach to the surface of our intestinal tract, ensuring that it will not be swept down to the large intestine when we next dine at our favorite restaurant. Depending on the species, a scolex may have sucker disks, hooklets, grooves, or even combinations of these adhesive structures with which to accomplish this important aspect of its life within us.

Tapeworms live at action central with respect to intercepting the food we eat: the upper small intestine (duodenum). I refer to these flatworms as nature's *gutless wonders*. I chose that term of endearment for a good reason: they have no gut tract; no mouth, no digestive system, no anus! In contrast, recall that the roundworm parasites (hookworm, ascaris, trichuris, dracunculus, filaria) all have functional gut tracts. This raises an obvious and interesting question: how do tapeworms obtain their nutrition? The answer is simple, but it took quite a while before scientists finally figured it out. It turns out that the surface of the entire tapeworm is covered with microscopic projections called *microvilli*. Millions of these minute projections serve to increase the surface area of each segment and provide a physical advantage for the efficient absorption of small nutrient molecules like amino acids, sugars, fats, nucleic acids, and vitamins. The presence of special proteins located in the actual membrane of each microvillus transports these metabolites across the surface of the worm, eventually getting them to the cellular machinery throughout the worm's tissues that converts our food back into large molecules specific to the worm's needs. In this way the tapeworm is able to grow from a barely visible immature organism into a parasitic behemoth in just over three months. Interestingly, under the heading "imitation is the highest form of flattery" (actually an example of parallel evolution), the surface of our own small intestinal cells has similar microvilli to those found covering the cestode. Absorption of nutrients is also quite similar in both cases. So it seems like our parasites are taking lessons from their hosts, and employing them for their own benefit. It's a fine example of evolutionary pressure enabling form to follow function when the biology of the situation demands similar solutions to similar problems.

Just as tapeworms have no digestive tracts, they also lack separate sexes. Each worm is both male and female, tied up in one big ribbon, so to speak. The segments are self-contained entities. They have their own nervous, excretory, and reproductive system (male and female organs), and muscles, too. The only thing each segment lacks is a head. In a very real sense, then, tapeworms can be considered a single giant colony of similar organisms, connected front to back. As one moves down the head to the last segment (several hundred in most large species), a complete series of developmental stages of the segments is encountered, with the least mature ones being found at the neck region near the scolex, and the fully mature segments with viable eggs being located at the opposite end. As each new segment is produced in the neck, it extends the colony's length. When full maturity is reached, the most mature (known as gravid) segment, now filled to the brim with viable eggs (hundreds per segment), breaks off from the colony and exits the host under its own power via the host's anus. This can happen anywhere and at any time. It usually happens when the host is asleep. The next morning, small, whitish, motile segments may be found on the very spot in the bed the infected individual just got up from. Taking them to a local physician's office is the most common way in which tapeworm infections are diagnosed.

But other situations, some quite dramatic, have also been recorded. Imagine for instance, standing on the subway platform at 168th Street and Broadway in New York, innocently waiting for the A train to pull in and whisk you off downtown, only to realize when you happen to glance downward that at your left foot, mere inches away from you, is a small, whitish, motile object making a beeline for the nearest exit. You wonder, "Is that a segment of a tapeworm? Is it mine?" Or how about dining at an upscale restaurant and, as the waiter approaches to take your order, you suddenly feel something quite unnerving going on in the lower right pant leg of the three-piece suit you just purchased for this very special occasion? It may sound like I just made up these two *tapeworm moments* to dramatize the fact that the worm segments can act on their own, but the truth is that these were two real events as re-lated to me by those who experienced them firsthand.

Once the gravid segment has left our body, the next step is rather problematic. The eggs inside it need to be ingested by another host. More precisely, it needs to infect another kind of host, depending on the species of worm in question. This can often prove difficult and, in parts of the world that are fully sanitized, next to impossible. That is why our tapeworms produce hundreds of eggs within each segment, in hopes that one will succeed. In nature, things are more, well, natural. Animals harboring tapeworms simply deposit a ribbon's worth of the parasite (several inches to a foot in length) on the ground with the feces during the act of defecation. The eggs are eventually released from each gravid segment, and a bit of luck gets them to their intermediate host animal. In these situations, earthworms and bird feet help to distribute the eggs from the original site of deposit over a wider area in the im-mediate vicinity, increasing the chances of the worm's ova being eaten by the correct host. The eggs are tough and can survive in moist soil for months. For several of the commonly occurring adult tapeworm species that infect humans, unsanitary conditions must prevail to allow for the spread of the infection. Since each case is unique, I will reserve the details of this part of the story for those examples when they arise.

Suffice it to say that after the eggs are eaten, the microscopic embryo inside each one hatches out into the lumen of the duodenum and im-mediately penetrates the wall of the small intestine. It enters the blood-stream and is passively carried to the heart, then out to all parts of the body. These larvae have no particular attraction for any given tissue, but skeletal muscle is abundant in mammals, so it becomes the most commonly infected tissue in most instances. Not coincidentally, it is this tissue that most of us prefer to eat, thus adding a note of behavioral reinforcement to the life cycle.

Once the immature tapeworm has lodged in a tissue, it encysts, de-veloping to the infectious stage for humans. Then it goes into a dormant phase and patiently awaits the arrival of the next stage in its life cycle. Humans must eat this larval stage to become infected with an adult tapeworm. Raw or undercooked muscle tissue is the most common source of at least three types of human cestode infections. The encysted immature worm is then released from its cyst wall by our digestive enzymes and is free to attach to the wall of the small intestine and begin to grow. In a little over three months, depending on the species, a fully grown, 20-foot-long adult tapeworm may occupy the better part of our upper small intestinal lumen. It can survive there for up to twenty years. Although many muscle cysts may have been eaten in the case of the beef or pork tapeworm, it is usual for only one tapeworm to become established. Perhaps intraspecific competition may come into play, giving the advantage to the individual worm that grew a hair faster than the others at the get-go, thus crowding out the rest of the wannabe parasites by garnering the lion's share of most of the available resources.

If there are no male and female tapeworms, then how do they produce eggs? Each segment has both sets of reproductive organs, so mating occurs between adjacent segments. The worm folds back on itself over and over again until all the segments contain sperm. That is how the ova are fertilized, yielding fully developed embryos within days. Mating can occur only between segments that have matured to a certain point in their development. This includes only those that occupy the middle of the colony of segments. As the eggs embryonate, the segments move toward the rear of the adult, eventually becoming gravid (with eggs), and it is at that point that they can detach from the parent worm and migrate out of the host.

Before traveling further on our journey into the wonderful world of tapeworms, let's take a moment to debunk a widely held belief regarding these much-maligned creatures. Tapeworm adults do not induce weight loss in their hosts, no matter how hard we have looked for evidence that would contradict that statement. So giving a tapeworm to someone who is overweight to help them lose weight simply does not work. That having been said, snake-oil dealers of the nineteenth and early twentieth centuries made fistfuls of money selling bogus tapeworm remedies out of the backs of their medicine wagons, and later on through mail-order catalogs. In fact, even if tapeworms did induce weight loss (which they don't), these flimflam men were selling capsules loaded up with the wrong end of the life cycle, the gravid segment. What's more, the segments were dead. Strike three! The only way an adult tapeworm could be acquired through the ingestion of a capsule is to have a live tapeworm head inside each one. The reader can now appreciate just how desperate some people can get, when they are willing to swallow a capsule guaranteed to give them a 20-foot-long parasite, but without the proper knowledge to judge whether that is a wise decision on their part. But on top of all that malarkey, to have the nerve to sell these unfortunate victims a stage of the parasite that has no chance whatsoever of developing into an adult worm is just plain fraud.

By the way, if you really want to acquire a parasite that is almost guaranteed to make you lose weight, get yourself hooked up with *Giardia lamblia*. This cute little protozoan induces malabsorption syndrome, which prevents fats from being absorbed. Steatorrhea is one of the side effects of the infection. What's *that*, you say? Steatorrhea moves the fat in the diet downstream to the eagerly waiting microbes living in the large intestine, where they then have a field day with that rarely seen commodity. That hubbub leads to foul-smelling flatulence, the by-product of anaerobic metabolism of fats, a most unpleasant consequence of giardiasis. So if its weight loss you want, then its weight loss you can have, but be warned. When you tire of endless bouts of loose stools, fatigue, loss of appetite, and being referred to by your former close circle of card-playing friends as the "outcast of *Poker Flatus*," and finally elect to take the cure (metronidazole), a rebound effect is the usual outcome, with a gain in body mass exceeding your original weight by about 10 percent. Sorry, no plowshare concept here. So much for using parasites as weight-control devices!

The Beef Tapeworm: *Taenia saginata*

The beef tapeworm is aptly named since we acquire the adult worm by eating raw or undercooked beef (fig. 6.1) that contains the infective tissue cyst (also known as the *cysticercus* stage). This is one of the most commonly occurring tapeworms infecting humans and is found in places where beef is commonly eaten rare or even raw, and where, on ranches, human feces containing the eggs of the adult worm routinely contaminates cattle feed. Mexico, Brazil, and most of West Africa fall into these two categories, and indeed, that is where most of the beef tapeworm infections occur. Most of western Europe and all of North America have very few reported cases of *T. saginata*. In the recent past, nearly twenty-five thousand head of cattle imported from Mexico each year were condemned owing to the presence of tissue cysts of *T. saginata*. Despite the low prevalence of infection in the United States, the U.S. Department of Agriculture still inspects for this parasite at most slaughterhouses by cutting deeply into the cheek muscles of the carcasses, then spreading the cut zone apart. Tissue cysts can easily be seen by using this method of inspection—they are white and are approximately the diameter of baby green peas. It is one of the few methods of inspection for parasites that does not require the use of a microscope. Carcasses testing positive are destroyed.

From a clinical perspective, the adult worm of *T. saginata* (fig. 6.2) does not cause any illness. I realize that it is hard to believe that a worm that size does nothing harmful to us, but such is the case. Try as clinicians have to pin a symptom or syndrome on this parasite, nothing comes up on the disease chart. The truth is that it just lives peacefully in the lumen of the small intestine and absorbs a small portion of the host's digested meal for its own needs. Maybe that is why it is so successful once inside us. Since it is not in close contact with our immune system, no host reactions occur that might make its life uncomfortable and cause it to leave. In fact, it is such an accommodating parasite

Taenia saginata

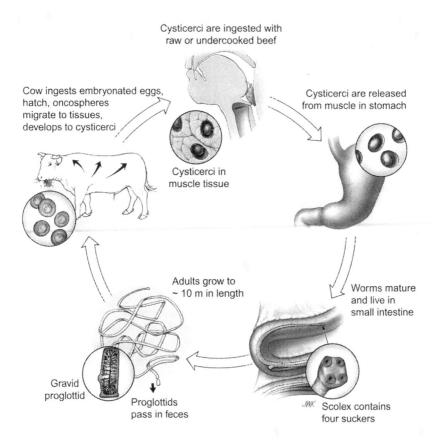

Cysticerci are ingested with raw or undercooked beef

Cow ingests embryonated eggs, hatch, oncospheres migrate to tissues, develops to cysticerci

Cysticerci are released from muscle in stomach

Cysticerci in muscle tissue

Adults grow to ~ 10 m in length

Worms mature and live in small intestine

Gravid proglottid

Proglottids pass in feces

JWK

Scolex contains four suckers

Figure 6.1 Life cycle of *Taenia saginata*. Illustration by John Karapelou. (*Source: Parasitic Diseases*, 4th ed., (c) Apple Tree Productions, LLC, New York)

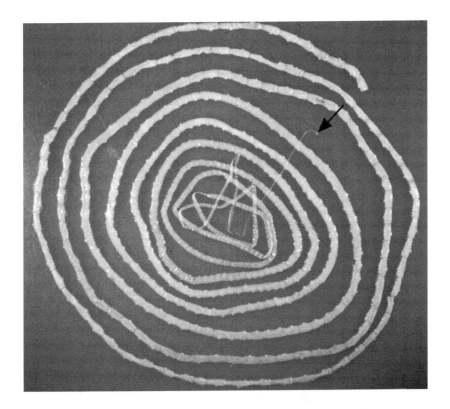

Figure 6.2 Adult *T. saginata*

that one investigator chose to infect himself with one so that he could always have segments available to study in his laboratory. Talk about dedication!

On the other hand, if you are a cow infected with the tissue cysts, you might experience any number of pathological effects, including neurological disorders related to these parasite larvae living in your brain tissue. Fortunately, if a human happens to swallow an egg of *T. saginata*, nothing happens. The egg (fig. 6.3) passes right through our digestive tract and on out with the feces. That is because we do not resemble a cow with respect to our intestinal arrangement. Cows

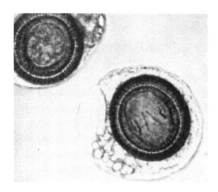

Figure 6.3 Egg of *T. saginata*

have several stomachs, and they use them first for digesting vegetable material (hay and such) and then for digesting the contents of the first stomach that feed on the vegetable material (bacteria and protozoans). The eggs of *T. saginata* are programmed to experience the full range of environments in both stomachs before hatching. They receive what are referred to as *environmental cues*. Since we do not "taste" the same as a cow from the perspective of the parasite, it cannot initiate its hatching mechanism once inside us. Later, when I discuss *Taenia solium*, the pork tapeworm, you'll see that this situation is no longer valid since pigs and people are quite similar with respect to their digestive tracts. The eggs of *T. solium* do hatch inside our digestive tract, and the migrating larvae cause a lot of trouble once they reach the brain or eye, but here I am getting a little ahead of myself.

If you happen to be diagnosed with an adult *T. saginata*, and you are not a famous parasitologist working on it in your suite of laboratories, and you really want to get rid of it, there is an excellent treatment—praziquantel. It inhibits part of the worm's defense mechanism that protects it from being digested by the host. When the drug works, the parasite is now vulnerable to attack by our enzymes that would ordinarily function to help digest our food. During treatment the worm simply dissolves and disappears out of sight (but probably not entirely out of mind).

The following vignette is one of the truly memorable episodes that occurred during my all-too-brief career as a parasitology diagnostic technician. It involved an unusual case of tapeworm infection and will serve to point up the fact that there is always something new under the sun, so be prepared for the unexpected! That particular incident went a long way toward convincing me that a career in parasitology was a great fit, that this special field of microbiology/ecology would provide me with all the intellectual challenges I could possibly handle, and, most important, that it was an early test of my limits for tolerating the truly disgusting.

Dr. Harold W. Brown was director of the parasitology diagnostic laboratory at Columbia/Presbyterian Hospital, located in upper Manhattan, from 1948 to 1970. He was one of the most generous, gracious, kind, and knowledgeable individuals I have ever met, and he was highly influential as a teacher in my formative years. In clinical circles in New York, he was a legend in the field of tropical medicine. I joined the lab as a greener-than-green technician in the summer of 1962. I was fresh out of college, and I had already completed a course in parasitology, but it was woefully inadequate for preparing me for the daily routine of a busy hospital diagnostic center. During my first week there, several incidents occurred that I could relate here that had nothing directly to do with parasites, but which would convince any reader that this particular subspecialty of medicine, parasitic diseases, covers the gamut from light comedy to dark drama. I've always been somewhat attracted to the darker side of things, and it's the drama that I now want to share with you.

The day in question began like every other day in the lab: specimens from the hospital delivered fresh daily (mostly stools, but some bloods and sputums too). As we started to record the samples in the data book,

Dr. Brown walked in from his office and asked me if I would like to accompany him to the wards that morning to look in on a patient of his. This was just day six into my new life. I was shocked and surprised, and also pleased. I recovered enough to say yes, and after I put on a fresh white lab coat and affixed my official-looking name tag over the upper left pocket to make me appear more physician-like, off we went into the bowels of Presbyterian Hospital in search of adventure. At least that is how I saw it.

We walked over to the hospital and took the elevator to the ward that had all the patients in the gastroenterology (GI) wing. On our way to see the patient, however, we had to first pass a gurney out in the hall with a teenage boy on it. He was hooked up to a respirator that was slowly pumping away. Most of the clinical staff in attendance (presumably some of them were the ones assigned to monitor the youth's condition) were bunched up on the other side of the aisle. A physician broke out of the huddled mass of white coats, went over to the gurney, checked the pulse of the boy, turned to the others, and shook his head slowly. As we walked closer, the respirator suddenly stopped its rhythmic cadence. We stopped too. The attending placed the sheet over the now deceased boy's face. I was frozen to the spot. I had never witnessed a death before, and Dr. Brown sensed this and ushered me off to his patient. On the way he surmised that the young man had most likely died of a drug overdose. The part of Manhattan we were located in was then infamous for its drug lords and gang wars. I am convinced it is also where Leonard Bernstein got some of his inspiration for *West Side Story*.

When we arrived at the patient's room, she was sitting up in bed reading the newspaper, but she was not happy. She looked about 50 years of age, with elegant, refined features, but obviously physically exhausted from her two days of hospitalization. All she was allowed to have for the entire stay was fluids. She had been given one drug by stomach tube (aspidium, an extract of the male fern plant) that was a nasty one with lots of side effects—nausea, vomiting, headache, dizziness. She had also been given another drug at the same time to prevent her from hurling up the aspidium, whose sole intent was to rid her of her adult tapeworm. She had spent the entire night without sleep, wanting to vomit but unable to do so.

She had returned two and a half months before from an extended vacation to the Middle East, which included a side trip to Turkey. While in Turkey the only rare meat she ate was beef since she believed that this was safe to do. This information had been very helpful in making a tentative diagnosis when she arrived several days earlier to the laboratory with a jar of whitish-yellow segments of what obviously was part of an adult tapeworm. She was then assigned a hospital room, and the day prior to our visit, Dr. Brown had instructed a GI resident to intubate her and administer the two drugs. The stage was now set for a tapeworm coming-out party! I forgot to mention that we had brought a plain white towel with us. When I asked what it was for, the good doctor just smiled and said, "We'll see, young man, we'll see." I was still thinking about that poor kid who had just passed away when Dr. Brown announced that I would be staying in the room with the patient, and as soon as she had a bowel movement, I was to wrap the bedpan with the towel and make my way back to the diagnostic laboratory immediately, with its contents intact. So I was to be the messenger. Cool!

Brown left us alone, waving good-bye as he headed out the door, and the patient resumed her perusal of the *New York Times* society pages. I sat there trying to imagine what was going on in her small intestines. Around half an hour later, she suddenly bolted up from her bed, wide-eyed, dashed off to the nearby bathroom, and within minutes emerged victorious with a grin on her face and a bedpan in hand, which she gladly relinquished to me as I held out my hands to take the baton, so to speak, in the tapeworm relays about to unfold.

Minutes later I arrived back at 600 Broadway, took the elevator to the fourth-floor laboratory, and entered the room, towel-covered bedpan in hand. Dr. Kathleen Hussey was in charge now. She was the one responsible for assembling and preparing all the microscope slides that the medical students had to sift over during their extensive course in parasitic diseases. A new tapeworm slide to be added to the collection was certainly a possibility that day. I gently handed the towel-covered bedpan over to the senior scientist, like it was a newly delivered baby, and everyone went to work. The unveiling was a triumph! The pan held a mass of ribbonlike worm that even Mario Batali would have been proud to serve, and there, sitting on the edge of the pan, was the head of the tapeworm. Bingo! No problem finding it, and a closer examination of it would prove which species of tapeworm it was, too. Suckers alone meant it was *T. saginata*, but suckers and hooklets were characteristic for *T. solium*. Unfortunately, in many cases the head of the worm goes missing, and other, less definitive characteristics related to the mature segments have to be brought into play for the diagnosis to be complete. Of course, today, the use of praziquantel eliminates any chance of ever recovering the head, or any other part of the worm. But there it was. I was enthralled, to say the least. The messenger had delivered the goods! One of the graduate students poured some warm saline solution into the pan to allow the worm the freedom to move about, greatly easing the job of separating the segments. As the worm unraveled, though, it suddenly became apparent that there was more to all this than first met the eye. Another tapeworm head was spotted, to whoops of surprise and joy! After it was placed in its own dish, yet another, and another, and another, until eight separate, complete worms, each about 5 feet in length, were culled and placed by themselves for all to see. Dr. Brown came back from clinical rounds just about then and asked what all the excitement was about. When he was told what had happened, he began to glow with pride and admiration. What a bonanza!

I recall that he gave me a pat on the back for a job well done. I was bursting with pride too. But there is a tag end to this story. The heads all were clearly *T. solium*-like, with suckers and hooklets, but the segments were obviously *T. saginata*-like. A new hybrid species? Wow. I was witness to tapeworm history in the making. The issue was never resolved, and to this day I have never heard of another case like this one. I assume the patient went home satisfied that she was no longer the home to eight alien invaders. I would have given anything to know if she changed her diet to well-done meats.

Controlling infection with *Taenia saginata* can be managed from several angles. Sanitation is the obvious most effective means of breaking the transmission cycle. But this is not an option in many regions throughout the world. Freezing the meat will kill the juvenile tissue cysts, but again, in less developed regions, electricity and refrigeration are often absent owing to the high cost of such luxuries. Meat inspection at slaughter is a third means by which infection with *T. saginata* can be prevented.

The Pork Tapeworm: *Taenia solium*

As they used to say on *Monty Python's Flying Circus*: "And now for something completely different." Well, at the outset, it might seem like this tapeworm is just like *T. saginata*, save for a few small tweaks. The mature worms are the same size, and both are clinically benign. They are both acquired by eating raw or undercooked meat (in this case, pork), and their transmission cycles are linked to the contamination of animal feed by human feces containing the eggs (fig. 6.4). All that is true, but as alluded to in the previous section, *Taenia solium* is much more closely in tune with our own intestinal physiology. Therein lies the trouble.

T. solium is found in many more places throughout the world than the beef tapeworm because pigs are far easier to raise than cows. Pigs will eat almost anything and therefore are more economical to keep, as they require far less space and achieve adulthood in just six months. In contrast, cattle take eighteen months to become fully grown. Because of their easygoing lifestyle, in many regions pigs live right alongside

Taenia solium

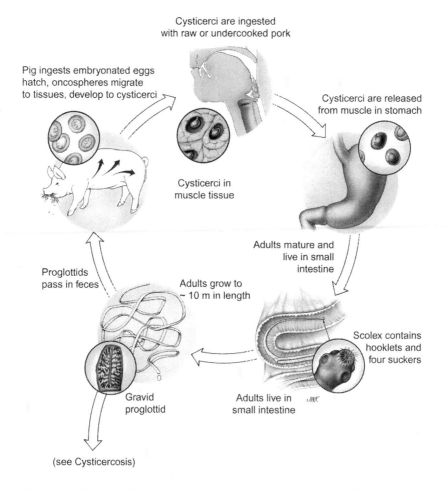

Cysticerci are ingested
with raw or undercooked pork

Pig ingests embryonated eggs
hatch, oncospheres migrate
to tissues, develop to cysticerci

Cysticerci are released
from muscle in stomach

Cysticerci in
muscle tissue

Adults mature and
live in small
intestine

Proglottids
pass in feces

Adults grow to
~ 10 m in length

Scolex contains
hooklets and
four suckers

Gravid
proglottid

Adults live in
small intestine

JWK

(see Cysticercosis)

Figure 6.4 Life cycle of *Taenia solium*. Illustration by John Karapelou. (*Source: Parasitic Diseases*, 5th ed., (c) Apple Tree Productions, LLC, New York)

people in what is referred to as a peri-domestic relationship. This sets the stage for a situation that has repeatedly brought sadness and even death to countless individuals living in communities spread throughout the tropical and subtropical world.

The disease is called neurocyticercosis. It is a common manifestation following ingestion of the eggs of *T. solium*. When a pig consumes these eggs, they hatch in the small intestine, and the embryo burrows into the intestinal tissue and eventually enters the bloodstream. It becomes distributed throughout the body and could lodge almost anywhere. But when it invades the central nervous system, not infrequently, bad things happen, like loss of motion in a limb, or loss of sight in an eye. Again, this aspect of the infection closely resembles that phase of the life cycle for *T. saginata* in the cow. The larva then develops to the infective stage for humans, and its volume, the size of a small green pea, affects the brain tissue much the same as any other space-filling lesion (e.g., tumors). The infected pig may develop symptoms depending on the area of brain affected. Pigs that behave strangely often end up at the slaughterhouse well before their time. When we eat raw or undercooked pork from such a pig, we can acquire the adult worm, thus completing the life cycle.

Now for the big difference between *T. saginata* and *T. solium*. If a person eats the eggs of *T. solium*, unlike the eggs of *T. saginata*, they will hatch in our small intestine and behave as if they had just been eaten by a pig. In other words, the parasite cannot distinguish between pigs and people. That is because a pig's digestive tract differs in significant ways from a cow's, but not from ours. The two stomachs of the cow are designed to harbor their microbial flora and fauna (remember that cows are ruminants), into which they feed plant material. The plants help grow bacteria and protozoans that then are swallowed, and it is this part of their digestive system that actually feeds the cow. It's a regular ecosystem of organisms designed to allow ruminants to feed on any plant material, regardless of where they live. Pigs, on the other hand, will eat anything. Just like us! They are omnivores, and their digestive tract is designed for this kind of eating pattern. How unfortunate for us. If many eggs are eaten, the majority of them might end up all over our body, with a few worms invading sensitive nervous tissue. If thousands of eggs are eaten, there is no doubt that many will end up in the brain, and with dire consequences.

Neurocysticercosis is a general term, so when the initial diagnosis is made, it is largely based on symptoms that the patient presents with in the neurological clinic. In many cases, it's a history of seizures or blindness in one eye or motion disorders of various kinds that brings the person to the hospital. On further investigation, the medical team in charge is obliged to test for all the major causes of neurological malfunctions— tumors, aneurisms, bacterial infections, and of course, infection with a juvenile tapeworm. Magnetic resonance imaging (MRI), a technique used to visualize the body in exquisite detail, will usually differentiate among these conditions, giving the physician a good chance of making the diagnosis with some certainty. Another follow-up test using the patient's serum can confirm the diagnosis if antibodies against this cestode are present.

What happens next is highly dependent on where the parasite has lodged. Often the tissue cysts are in places that can be approached surgically, but there are times when the organisms cannot be removed without risking the life of the patient. In these cases, drugs are used to kill the parasites. When they die, the worms relinquish their hold on our immune system, and an intense inflammation is the result. Many patients do not even know that they have been infected until the parasites begin to die—around ten years after the tissue cysts form. Treatment involves giving steroids to manage the inflammation and swelling of surrounding tissues after the worm dies.

Dr. Robert Desowitz, a renowned parasitologist and longtime friend, wrote a wonderful book with the whimsically enigmatic title *New Guinea Tapeworms and Jewish Grandmothers*. In it he described medical anthropological aspects of tapeworm infections, including a huge outbreak of neurocysticercosis in Irian Jaya, New Guinea, caused by *T. solium*. I highly recommend it as an entertaining and accurate account of how the government of Indonesia made a gift of pigs to several tribes living in Irian Jaya to smooth the away for a reunification effort between Bali and Irian Jaya. Tragically, those pigs were carrying the tissue cysts of *T. solium*. The rest of the story recounts what happened next. Pigs are used in many New Guinea cultures as a show of strength. Annual feasts end the yearlong vying for leadership among the various tribes. Whoever supplies the most pigs for the feast is declared the winner. All the animals are eaten, and they are never cooked well enough to kill the parasites. In the past Irian Jaya did not have *T. solium* infection so that was never a problem. That is, until the fatal pigs arrived. Around two to five years later, a number of burn victims suddenly began showing up in regional clinics. All had suffered from seizures and fallen into their home fires. Seizures had never been observed before in those populations. Furthermore, routine parasitological surveys using stool samples identified taenia eggs in those same populations for the first time. Given these facts, epidemiologists were able to show that there was a direct correlation between the introduction of diseased pigs and the onset of neurocysticercosis. Unfortunately the problem still exists today, as the residents of that part of New Guinea cannot escape their cultural habits to prevent the spread of the infection.

Another unusual outbreak of neurocysticercosis occurred in 1992, in a most unlikely neighborhood: the Ocean Parkway section of Brooklyn, New York. What made the epidemic particularly odd was the fact that that area of New York is settled by a predominantly sephardic Jewish population. Pork is not one of the choices on their menus, so how did four unlucky residents (all unrelated to one another), and three more who were also most likely infected, acquire their infections? The Centers for Disease Control and Prevention was called in to investigate after the epidemic had subsided. Epidemiological Intelligence Service

(EIS) physicians queried everyone involved, and it took weeks of careful sifting of data before the penny dropped. It seems that one particular household had hired a domestic worker who came from a small town in northern Mexico. She was so competent that the word spread quickly into the neighborhood as to her hard work ethic and pleasant personality. Neighbors began to inquire as to the availability of others like her from the same town. Soon there were several more Mexican workers from that town employed in the same area of Brooklyn. In the end, six women had brought their adult tapeworms with them and had subsequently exposed the residents of Ocean Parkway to the infectious eggs. Later it was shown by careful whole-body examinations using cotton swabs and microscopy that those individuals who harbored adult *T. solium* had eggs in many unsuspected places—under fingernails, between fingers, behind ears. These data were the basis for revising our view of how a person harboring an adult worm might transfer eggs to others in their immediate environment. It is estimated that around 25 percent of all those with adult *T. solium* are infected with tissue cysts as well.

Treatment for the adult worm infection is the same as for *T. saginata*, praziquantel. Prevention is also similar, but as long as pigs and people continue to occupy the same domestic environment, *T. solium* infection will persist.

The Fish Tapeworm: *Diphyllobothrium latum*

The winner of the Longest Parasite to Infect Humans Contest, by a wide margin, is the fish tapeworm, *Diphyllobothrium latum*. Specimens have been recorded up to 30 feet in length. It is called the fish tapeworm because we acquire the infection by eating the larval stage that infects predatory freshwater fish, such as northern pike and walleyed pike, yellow perch, salmon, and trout—but only if we eat raw or undercooked infected fish preparations. Some estimates suggest that there are as many as twenty million people worldwide harboring this parasite. Unlike *T. saginata* and *T. solium*, other animals can also acquire the adult worm of *D. latum*—bears, dogs, cats, seals, weasels, otters. Only a few cuisines, in which fish is either served raw (e.g., gravlox) or smoked, but not cooked, favor the transmission of this tapeworm to humans. Sushi made from these fish species is certainly not recommended. Most sushi is made from saltwater fishes, so there is no risk of acquiring this infection by eating these fish varieties. But smoked and raw salmon can be ordered at many sushi bars. Today, most salmon is farm-raised, lowering the risk of infection with *D. latum*. In past years, *D. latum* was routinely acquired during the preparation of gefilte fish, during the act of tasting the raw preparation to check for the correct balance of spices before actually cooking (see *New Guinea Tapeworms and Jewish Grandmothers* for an entertaining and somewhat lighthearted account of how *D. latum* is transmitted to the young women of some Jewish families). The Upper West Side of Manhattan was home to many Jewish families in the 1950–1970s. During my brief tenure as a technician, our laboratory diagnosed and Dr. Brown then treated many cases of this parasite found in the local residents of Washington Heights. I have kept a souvenir of that era, a beautifully preserved specimen of an intact adult *D. latum*, which I proudly display on my desk. It's a sure-fire conversation starter!

The fish purchased for making gefilte fish by those living on the East Coast came mostly from the Great Lakes. In the past, raw, untreated municipal sewage was indiscriminately dumped into most of those freshwater bodies. A large influx of Scandinavians to the Midwest during the 1900s, coupled with these unsanitary practices, established and maintained the life cycle of *D. latum* in all the Great Lakes, except for Lake Superior. Local residents living along those lakes were equally at risk of acquiring *D. latum*. It was generally acknowledged that the best fishing was right through the dump hole of the outhouses located at the end of their docks.

Countries that experienced high rates of infection in the past included Finland, Sweden, Chile, most West African countries, the Czech Republic, Canada, Russia, Japan, and Brazil. The situation today is much improved in many parts of the world, in particular the Scandinavian and Baltic countries, and their rates of infection with *D. latum* have plummeted. But elimination of this tapeworm is not just dependent on improving the safe disposal of human waste. Animal reservoirs play a major role in some areas, particularly in the Pacific Northwest, where wild salmon and brown bears still thrive.

The way in which *D. latum* gets from one host to the next is truly a fantastic voyage (fig. 6.5). It involves a series of events in which several different intermediate hosts and one definitive host must all play their role in the parasite's life cycle if it is to succeed, and succeed it does. This worm has developed a complex relationship with the rest of the natural world that would impress anyone with an appreciation for the intricacies of life on Earth.

The adult worm's biology is similar to the taenias', with several exceptions. The head of the adult does not have any hooks or suckers with which to hold onto us. Rather, it has two long grooves on either side of its torpedo-shaped head that it spreads apart on contact with the surface of our small intestines. By drawing back on itself, it creates a partial vacuum and remains comfortably in place as the result. Also unlike the taenias, infection with the adult of *D. latum* can eventually lead to a disease called megaloblastic anemia, caused by vitamin B12 deficiency. The longevity of the adult worm is well over ten years, and during that time it absorbs all our dietary vitamin B12. About two to three years after the initial infection, as a result of the sequestration of this essential compound in worm tissue rather than ours, we begin to experience the effects of the vitamin deficiency. That is because our bodies store surplus vitamin B12 in our liver, and most of us have at least a two-year reserve. It is still unresolved as to why the worm is

Diphyllobothrium latum

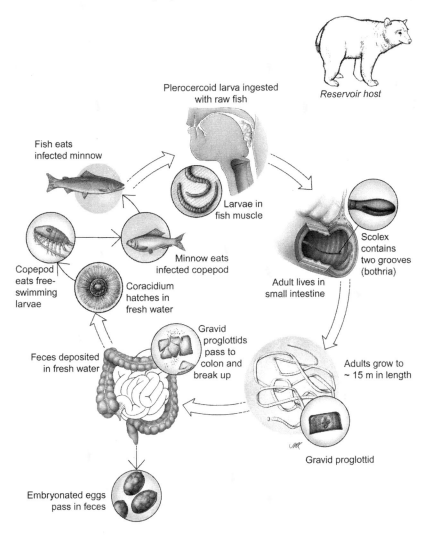

Reservoir host

Plerocercoid larva ingested with raw fish

Fish eats infected minnow

Larvae in fish muscle

Scolex contains two grooves (bothria)

Minnow eats infected copepod

Adult lives in small intestine

Copepod eats free-swimming larvae

Coracidium hatches in fresh water

Gravid proglottids pass to colon and break up

Feces deposited in fresh water

Adults grow to ~ 15 m in length

Gravid proglottid

Embryonated eggs pass in feces

Figure 6.5 Life cycle of *Diphyllobothrium latum*. Illustration by John Karapelou. (*Source: Parasitic Diseases*, 5th ed., (c) Apple Tree Productions, LLC, New York)

hoarding this essential compound, but it is likely using it for its own DNA metabolism, to support its prodigious egg production.

Segments found at the tail end contain hundreds of fertilized, unembryonated eggs. These gravid segments periodically shed their eggs through the uterine pore and the eggs then exit the host with the feces. To develop and hatch, the eggs must first be deposited in a freshwater environment, preferably one that does not move (e.g., a lake). Occasionally, gravid segments break off and exit the host. Eggs are then released into the aquatic environment after the segments disintegrate. Slow-moving rivers can also support the aquatic cycle of this parasite, but lakes are better because of what must happen next. Once in water, the embryos begin to develop, and within a little more than a week, they hatch out from their shells and begin to swim.

Each larva (called a *coracidium*) is covered with cilia, hairlike structures that act like little oars and help the organism move about in the water column. *D. latum* is now a free-living entity; quite unusual for a parasite (recall from chapter 4 that the schistosomes have a free-swimming stage, too). They are then eaten by planktonic crustaceans (e.g., water fleas, cyclops). These barely visible creatures abound in most freshwater lakes throughout the world and are distant cousins of more familiar life forms such as microscopic shrimp and crabs, which have voracious appetites and will eat nearly anything smaller than themselves, particularly things that move, including the coracidium of *D. latum*. Big mistake! Instead of being digested, the larval parasite burrows through the microcrustacean's digestive tract and parasitizes it, developing to the next stage (the *procercoid* stage) in its complicated life cycle.

The food web in freshwater lakes involves several levels and hundreds of participants, from algae to crustaceans, minnows, and predatory fish. Minnows are filter-feeders and as such consume planktonic organisms, including the infected microcrustaceans. The water flea or cyclops is digested in the minnow's small intestine, freeing the larva of the tapeworm. The parasite then penetrates the gut tract of the minnow and comes to lodge between the muscle fibers. It develops further into the infectious stage for humans (the *pleurocercoid*). But there is a catch (pun intended). We do not eat a lot of minnows, nor do any of its other reservoir hosts. So for the parasite to gain access to our food supply (the predatory fish), it is transferred to them when these larger fish feed on minnows, some of which are harboring the parasite. This stage can locate to the same part of the predator fish as it was in the minnows. Now all we have to do is catch one and eat its flesh raw or undercooked to acquire the adult tapeworm. The white, threadlike pleurocercoid stage is often mistaken for a stray muscle fiber, even when experienced parasitologists examine fish carcasses looking for the parasite. It is also quite easy to overlook when preparing gefilte fish!

Again, as with the other adult tapeworm infections described above, treatment is with praziquantel. Improved sanitation has practically eliminated this infection from most developed countries.

The Dog Tapeworm: *Echinococcus granulosus*

If *Diphyllobothrium latum* is the "Jeff" of tapeworms, then *Echinococcus granulosus* is definitely the "Mutt" (fig. 6.6). This adult worm is a mere three segments long and measures only 3–6 millimeters in length when full grown (fig. 6.7). What's more, *E. granulosus* infects dogs, not humans. The intermediate host is usually a sheep or goat. So, what's all the fuss about? It is because the larva of this little tapeworm cannot distinguish between sheep, goats, and humans. Infection with the larva of *E. granulosus* produces a space-filling object called the *hydatid cyst* that can grow up to the size of a grapefruit!

The dog tapeworm is found wherever sheep, goats, cattle, camels, or buffaloes are raised—Argentina, Chile, Australia, New Zealand, most European countries, Turkey, Cyprus, South Africa, India, Iran, and the southwestern United States. It used to be endemic in Iceland, but a re-

Echinococcus granulosus

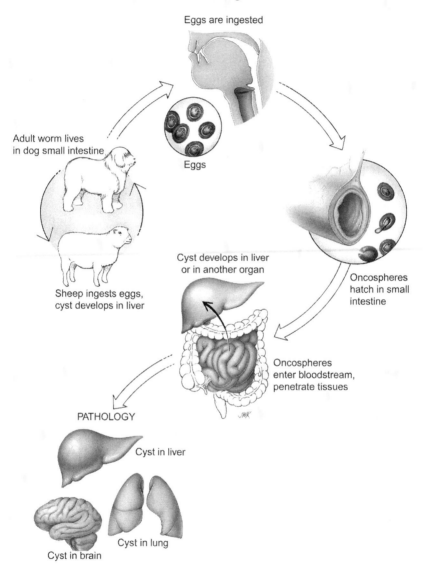

Eggs are ingested

Adult worm lives in dog small intestine

Eggs

Sheep ingests eggs, cyst develops in liver

Cyst develops in liver or in another organ

Oncospheres hatch in small intestine

Oncospheres enter bloodstream, penetrate tissues

PATHOLOGY

Cyst in liver

Cyst in lung

Cyst in brain

Figure 6.6 Life cycle of *Echinococcus granulosus*. Illustration by John Karapelou. (*Source: Parasitic Diseases*, 5th ed., (c) Apple Tree Productions, LLC, New York)

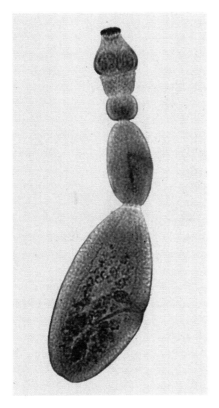

Figure 6.7 Adult of *E. granulosus*

lentless public health education program and prevention of exposure of dogs to offal from slaughtered infected animals has all but completely eradicated it. Another well-documented, successful multipronged approach to controlling the infection on the island of Cyprus, initiated after 1970, has largely eliminated this infection there, as well.

Dogs (mostly sheepdogs) can accommodate many thousands of these small tapeworms, making their feces highly dangerous to us. That is because millions of infectious eggs could easily be produced each day by just one dog. The eggs remain viable in the environment, depending on local conditions. High humidity and warm temperatures favor their long-term survival. It only takes a single egg to produce a hydatid cyst. That is why public health departments in the countries with high prevalence have active control programs to limit the spread of this potentially fatal infection.

The life cycle is similar to that of the taenias. The adult worm lives in the small intestine of the dog. The last of its three segments contains the eggs and is passed in the feces periodically, as a new segment is added. The eggs in the gravid segment are fully embryonated and ready to infect the next host. Grazing animals that share their pastures with dogs are at highest risk for becoming infected with the hydatid cyst. When an egg is eaten, the embryo inside hatches out in the small intestine. The larval tapeworm burrows through the intestinal wall, enters the bloodstream, and, in the majority of cases, the circulation takes it to the liver, where it begins to differentiate into the hydatid cyst (fig. 6.8). Other organs have been recorded in which a hydatid cyst has formed—brain, lung, bone.

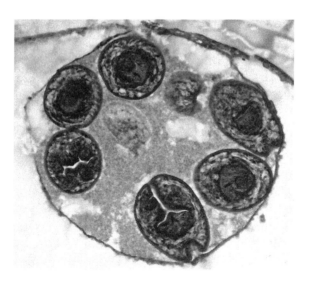

Figure 6.8 Portion of a hydatid cyst of *E. granulosus*

E. granulosus is a master at producing stem cells. As the cyst forms, a thick layer of material is produced by the developing parasite that encases all the newly synthesized microscopic tapeworm heads, called protoscolices. If we were to imagine how the parasite is arranged inside the cyst, we could compare a mature hydatid cyst to a set of nested Russian dolls as a good analogy. The cyst takes several months to attain full growth and is filled with a nonviscous fluid that helps it maintain its shape throughout its development. Should the cyst rupture for any reason (usually a trauma injury), the contents will spill out and disseminate cells from the cyst into the damaged area. In human cases, this has proven fatal.

So how does the parasite infect a dog once the hydatid cyst has formed in the intermediate host? It's dependent on patterns of human cultural behavior that determine the fate of the cysts once they are detected in a domestic herd of animals at the time of slaughter. In many nomadic cultures, for example, sheep, goat, and camel livers that are affected are customarily given to the herder's dogs. This results in millions of protoscolices developing to adult tapeworms in the small intestine of the dog. Surprisingly, this human behavior is difficult to disrupt by interventions that are well intentioned, but that originate from government or public health agencies. Other types of government programs are better received, such as teams of public health workers who are charged with periodically treating all domestic dogs with praziquantel. Apparently, in addition to Iceland and Cyprus, this approach has worked well in rural parts of the American Southwest on Native American reservations.

A Bizarre Parasite: *Echinococcus multilocularis*

Echinococcus multilocularis apparently loves reading Stephen King novels, because its biology, at least after it infects us, is truly off-the-wall scary. Well, to be fair, it's not *so* unusual a parasite if you happen to be a fox, or a dog, or a rodent. But if you are a human, there is no other infection equivalent to it anywhere on Earth.

The adult worm is a member of the "Mutt" tapeworm family, just like its close cousin, *E. granulosus*. This three-segmented cestode lives attached to the wall of the small intestine of canids (foxes, dogs, wolves, coyotes), while the larval form is found in the liver tissue of various rodents. The distribution of *E. multilocularis* is global, being found in both the Northern and Southern Hemispheres. It is an introduced species in many places, including New Zealand. It probably got there with the foxes that were brought in for the sport of hunting, or from hunting dogs gone wild.

The biology of this tapeworm follows the usual pattern: adult worm, to egg, to cyst in an intermediate host (rodents), and back to its definite hosts, as adult worms, after they eat the tissue cyst. The tissue cyst is unique, comprising an interconnected series of membranous partitions, inside of which are the immature tapeworm heads. In tissue sections its appearance is that of a honeycomb with several heads in each compartment. Each cyst may contain thousands of parasites. When the infected rodent is eaten, each tapeworm head evaginates and attaches to the intestinal wall of the canid's small intestine. Shortly thereafter the adult worm begins to shed gravid segments laden with infectious eggs into the feces. Each egg can initiate infection in a rodent, the result of which is the formation of the multilocular tissue cyst. So far, so good. Nothing too bizarre here. But as they say on the late-night TV ads, "Wait. There's more!" What if *we* were to eat an egg? What then?

Now the weirdness starts. First off, the larva of *E. multilocularis* is unable to fully recognize us. This clearly distinguishes it from *E. granulosus*. Perhaps we should derive some small comfort from the fact that we are not rodentlike, at least as seen through the eyes of a parasite. Unfortunately the parasite discovers that we are not the right host *after* it has hatched, penetrated our gut tract, and traveled to our liver. Nevertheless, it still tries to make the best out of a bad situation. Recall that this group of tapeworms is renowned for producing stem cells. In this unusual situation, the stem cells begin to grow in our liver as sheets of thin membranes. No identifiable tapeworm structures are produced, just membranes. OK, you say, so what? No harm, no foul. Well, these membranes can bisect the liver many times over, eventually disrupting all the functions of this crucial organ. Ultimately, many years down the road, it proves fatal. There have not been too many recorded cases of this highly unusual condition, but each one has captured the attention of parasitologists.

It appears that the geographic range of *E. multilocularis* in North America is moving south from its original subarctic habitat and is now found throughout the upper Midwest of the United States. Perhaps global climate change has had an effect on its distribution. Only time will tell if we have a more serious situation to contend with over the next twenty to fifty years, as this tapeworm continues to expand its range to more populated areas.

Plowshare Concept

If we could put any of the tapeworms to work for us, I would choose the juvenile (cyst) stage of *Taenia solium*. The cyst of this parasite has the ability to shut down our immune system from mounting an attack on it and can maintain that suppressive activity for up to ten years. This is strong indirect evidence suggesting that it is capable of secreting a powerful immunosuppressive agent. Such agents would be of great value in many medical situations, particularly for organ transplants. Isolating the active ingredient in this tapeworm's arsenal of survival tactics could be approached by cloning all its genes (so-called shotgun cloning) into a mammalian cell culture that we could then use to search for the immunosuppressive agent. There are lots of well-developed biological assay systems for just such a research project. Its likely that we could find the desired molecule if the genes responsible for producing it are not harmful to the cells into which they were cloned. After that, it's a matter of identifying the worm substance and perhaps even being able to synthesize it in the laboratory. So even though these seemingly primitive critters appear to be only interested in themselves, we can begin to imagine how it is not only about them, but about us, too! In the end, our job is to find ways of getting them to share their secrets with us. Put another way, we have to learn how to speak tapewormese!

7 ALL'S WELL THAT ENDS WELLS

Dracunculus medinensis

The caduceus, classically depicted as a pair of snakes wrapped around a two-winged staff, is a familiar iconic symbol for all things having to do with medicine (fig. 7.1). Even the World Health Organization has incorporated a modified version (one snake and a staff, no wings) as part of its blue and white logo. The origins of the modern caduceus can be traced back as far as the ancient Greeks and their richly expressed mythologies. Yet its original form and purpose(s) are considered to be much older, perhaps even dating back to prehistoric times. Over the ages, its basic shape and dimensions have been modified, perhaps to suit the needs of the moment. But here it is, often embroidered in bright red on the lapel of white coats of practicing physicians and residents alike, worn proudly and as prominently displayed as the silver metal wings of any airline pilot. It has become a sign of authority. Nurse Ratched (*One Flew Over the Cuckoo's Nest*) brazenly wore a brilliant red version of it on her highly starched, lily-white uniform; a kind of in-your-face caduceus. Curious, that, since most doctors or nurses rarely if ever get the opportunity to treat snakebites! If it's not actually some kind of talisman to ward off hordes of aggressive serpents, then what the heck is it? In fact, are those really snakes, and is that an actual staff or is it a sword, or maybe even a plain old stick? What *does* this enigmatic

Figure 7.1 Caduceus

symbol stand for? Some brief historical vignettes might help set the stage for our next parasite story.

Today many people, particularly those of the Jewish faith, believe that the original figure of a man leaning on a large staff, around which is wound an enormous, serpiginous reptile, represents Moses at the time of the Exodus. He was supposedly acting on the word of God, taking his followers into the wilderness to the east, a very dry and inhospitable desert environment. One can readily understand how easy it might be for anyone following that much-revered figure to lose track of the fact that God instructed him to go in that seemingly hopeless direction. No water, no real food (unleavened, tasteless bread made in haste), and, worse yet, nothing in sight to give them hope of the "promised land" of milk and honey. Whenever I think of this familiar biblical story, I cannot help but imagine a comedic monologue delivered deadpan by some gifted entertainer like Bob Newhart, Lewis Black, or Bill Cosby, re-creating this scene on stage and injecting all sorts of ironic humor into it. But in reality it must have seemed to many of the participants (if it actually *did* take place) like some kind of forced mass-suicide march. Those among the escapees who lost their confidence early on during the trek spoke out against God's plan to save them. In response, God became angry (a very ungodly but a very human thing to do) and unleashed a plague of fiery serpents on the multitude that spared the faithful and only bit the ones that got out of line. Moses, horrified at the wrath of such a God, took pity on the smitten and asked God to back off. God caved in (another very human characteristic) and empowered Moses to save them by having each envenomed mutineer touch Moses's bronze serpent, the one he somehow fashioned and then managed to wrap around his wooden staff (in record time, too!). It must have been very slow-acting venom. Hard to imagine what kind of snake that could be, too, unless of course it was not a snake.

The Old Testament relates the story thusly:

> Then the LORD sent fiery serpents among the people, and they bit the people, so that many people of Israel died. And the people came to Moses and said, "We have sinned, for we have spoken against the LORD and against you. Pray to the LORD, that he take away the serpents from us." So Moses prayed for the people. And the LORD said to Moses, "Make a fiery serpent and set it on a pole, and everyone who is bitten, when he sees it, shall live." Moses made a bronze serpent and set it on a pole. And if a serpent bit anyone, he would look at the bronze serpent and live.

It all sounds a bit farfetched, don't you think? Personally I do not adhere verbatim to any ancient historical accounting, since those records have been muddied many times over by several translations from the original text to Greek, then Latin, then English. Most archeologists also agree with my take on the distant past's literature. In fact, none of them can find any evidence whatsoever regarding the Exodus. None of the thousands of unearthed tombs near Cairo or anywhere else in Egypt, for that matter, have contained any human remains other than Egyptians. Furthermore, even though we have access to some of the original documents of the Old Testament in the form of the Dead Sea Scrolls, we have no idea what the original Hebrew words actually meant in many cases, since we cannot reconstruct the social life of the time and listen in on the everyday conversations. It is only by inference that we can glean the approximate gist of anything in the distant past, be it Egyptian or Mayan hieroglyphs, or the language of the Dead Sea Scrolls. I do not claim to be an etymologist, nor am I a gifted linguist, but I have attended enough meetings organized by medical historians and heard them waffling among themselves about the meaning of this or that to know that we cannot take at face value anything that refers to the word of God, or Moses, for that matter. For some perverse reason, the Mel Brooks take on how we got the Ten Commandments comes to mind here, but that's just me.

Science seeks alternate explanations for these familiar religious beliefs in the hope that the actual truth may lie in the interpretation of something more interesting that really happened or existed at the time of the writings. Explaining the natural world must have been hellish for anyone without the foundations of modern science to help them out. Many ancient civilizations adhered to a multigod universe, using this as the mechanism to explain everyday phenomena such as a solar eclipse, outbreaks of disease, and adverse natural events (floods, droughts, etc.). Later on, the amalgamation of all these gods into a single deity by the Hebrews made it a lot easier to deal with all the tumult of living on Earth. Attributing all the unexplainable things under the rubric "acts of God" made society bearable, and dare I say more rational. Whether any of the remarkable things that we take for granted as having been done by the hand of God actually happened is a matter of conjecture and will never be resolved by any approach, scientific or otherwise. They will remain in the realm of a *belief* rather that a *proof*. Nonetheless, many believe that the caduceus icon is based on a real thing rather than a Moses intervention strategy to save as many Israelites as possible for the new colony.

Finding evidence that hand-carved, metal-cast, or pictogram versions of snakes wrapped around sticks existed long before the time of Moses would remove a bit of the luster from that well-known story and would add a touch of realism to its origins, and here we are in luck. The ancient Greeks knew about this symbol long before the Romans. In legend, Hermes (Mercury to the Romans) was given a magical wand (kerykeion) by Apollo in exchange for Hermes's lyre, which he had fashioned from the shell of a tortoise (seems like a fair trade to me). Hermes then wrapped two long ribbons around the wand (really a staff) and thereafter was never seen without it in his hand. As the myth goes, one day Hermes happened on two snakes locked in mortal combat. Using the magical powers of his newfound toy, he easily pried the two opponents apart and brought peace to the valley. As a commemorative of that event, he discarded the two bits of ribbon and insinuated the two snakes on the wand instead. Thus was born the first of many versions of the modern caduceus. In this original iteration, it stood for tranquility, having the two snakes now placidly displayed on his staff. I am sure I have just completely obliterated the flavor and richness of the original prose of that Greek myth, but the essence of it is clear enough. The medical version of the caduceus was now well on its way to becoming the icon we all recognize today. But what about before that (ca. 2612 B.C.E.)? Were there even older examples of representations of sticks and serpents, or is that the end of our story? Apparently there are older versions from the Middle East (Syria) and India. In these cases, they were definitely not part of the Western view of *caduceusism*. Then what were they depicting, and could they have been adopted by more recent cultures to suit different purposes?

It is now time to reveal the parasite part of our story. There is a particular long-lived roundworm that just might fit the bill for the snake part of the story, and a simple stick is what we (at least we parasitologists) now think makes up the other half of this iconic pair. The disease is called dracunculiasis, and it's a disabling widespread infection found throughout the rural parts of Africa, the Middle East, India, Bangladesh, and some parts of Latin America. It is associated with semiarid environments, and also places where safe drinking water is a rare commodity. Aha! you say. Aha! I say along with you. The worm is named *Dracunculus medinensis.* Draco translates to "dragon" or snakelike, and the *medinensis* part of its moniker refers to the modern version of the caduceus. How this worm behaves once inside us, and how we react to its presence, is the heart of our story regarding the origins of this symbol of healing, and the one most likely to be the closest to the truth. The worm is known by other names too. One of them is the *fiery serpent of the Israelites.*

All ancient civilizations whose origins were in the Middle East and Africa knew about this parasite, even though they had no clue as to how it was transmitted. One thing was certain, however. They all knew how to get rid of it once they became infected. The worm is long, and its head is located in the lower extremities. The adult parasite causes a blister to form in the host at the location of the worm's head, and it causes the victim much pain and discomfort. Submerging the foot with the parasite in it in water allows the blister to break open, relieving the pain, and at the same time exposes the head of the worm (fig. 7.2). The cure is to grab the head of the beast and wrap it onto a thin stick. But extreme care must be taken with this delicate procedure. Turn the stick at exactly the same pace that the adult worm relinquishes its hold. If the victim becomes impatient and rushes things, the parasite will break. Imagine yourself trying to pull an earthworm out of its burrow. Pull too hard and you would have to find another half a worm if you want to go fishing. In the case of dracunculus, breaking this worm proves disastrous. The contents of the severed roundworm elicits a very painful, massive hyperallergic reaction that could cost an individual the use

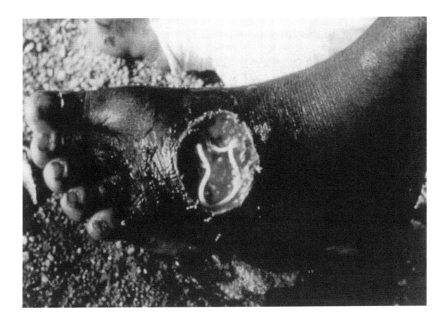

Figure 7.2 Wound caused by *Dracunculus medinensis*

of his or her infected foot and leaves an ugly scar down the entire side of the infected leg (fiery serpent indeed!). What's more, the open sore usually becomes secondarily infected with bacteria, complicating even further the disease caused by the worm. If done properly, the whole process takes some two to three weeks to carry out. All the while, the victim sits at home and endures this parasitic insult while trying to carry out all the other things in his or her life that constitute routine activities.

The image of the single stick and a lone, snakelike object wound around it fits perfectly with this all-to-common scenario; one that must have been repeated over and over again throughout the many occupied rural tropical zones of the world. It's no wonder then that images of this simple remedy would become cultural icons, depicted on temple walls, made into jewelry or wooden objects, and described in various ways in the religious and medical texts. Today dracunculosis is a disease mostly associated with poor and disadvantaged people. The good news is that it may join those other few infections (smallpox, yaws, polio, river blindness) that soon could become extinct at the hands of us humans. Like these other parasites, *D. medinensis* has no significant reservoir hosts, making programs aimed at its elimination at least theoretically possible.

The infection is acquired by the simple act of drinking water (fig. 7.3). We need 2.3 liters of it every day, and it better be pathogen free if we are to live out our lives in relatively good health. Unfortunately most of the world does not have access to safe drinking water, and many do not have access to a reliable source of freshwater, period. The result is that a plethora of water-borne diseases, mostly diarrhea-causing microbes, constantly affect the daily lives of billions of people worldwide, resulting in high mortality rates, particularly among the very young. Although *D. medinensis* is an infection associated with polluted freshwater, it does not induce diarrhea. Yet in the vast majority of rural settings where *D. medinensis* is found, these two clinical conditions (diarrhea and dracunculosis) exist side by side. The worm is associated with a very unusual clinical pathological feature—clearly visible circular scars, usually found on the lower extremities. Many international control programs initiated in the 1980s were aimed at identifying unsafe sources of drinking water and were designed to take advantage of this visible feature, making it relatively straightforward for a health professional doing potable water surveys in rural environments to identify a victim of *D. medinensis* and draw the logical conclusion that the water was still unsafe to drink. No complicated microbial culture tests and fancy laboratory equipment needed here. Just look for the tell-tale scars and record the result.

In 1990 there were an estimated ten million people infected in Africa alone. Nineteen countries were involved, with Guinea in West Africa being the epicenter for the infection and thus its logical poster child.

Dracunculus medinensis

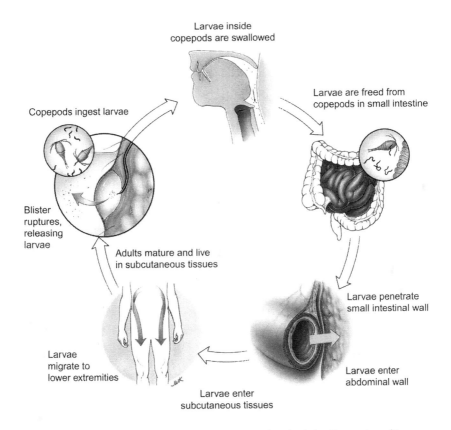

Larvae inside copepods are swallowed

Larvae are freed from copepods in small intestine

Copepods ingest larvae

Blister ruptures, releasing larvae

Adults mature and live in subcutaneous tissues

Larvae penetrate small intestinal wall

Larvae migrate to lower extremities

Larvae enter abdominal wall

Larvae enter subcutaneous tissues

Figure 7.3 Life cycle of *D. medinensis*. Illustration by John Karapelou. (*Source: Parasitic Diseases*, 5th ed., (c) Apple Tree Productions, LLC, New York)

Another name for dracunculosis is Guinea worm disease. Today the number of cases has been dramatically reduced to less than ten thousand. Many countries are now free of the infection altogether. All this has come about as the result of a concerted effort on the part of numerous international health agencies and nongovernmental organizations such as the Carter Center in Atlanta. Even the Peace Corps was involved at one point. I'll come back to this control program later, but first let's learn a bit more about the biology of this parasite.

As mentioned, the adult worm is quite large, measuring up to 100 cm in length by 1.5 mm in width. There are many related species, but parasitologists who study the filarial group of worms generally agree that only one, *Dracunculus medinensis*, infects humans. Furthermore, its also believed that there are no significant animal reservoirs for the human infection. The adult worm lives embedded in the subcutaneous tissues, just under the dermis (outer skin), for up to a year. Its head is pointed down and is usually located at the level of the ankle or the top of the foot. But many exceptions to this general pattern have been recorded in which the adult worm has wrapped itself around the Achilles tendon, then died and calcified, rendering the foot immobile, thus crippling its victim. In other instances, the head of the worm has emerged from the bottom of the foot, or out of the upper chest area. Regardless of its anatomic location, its mission is to give birth to live larvae (like the filarial worms and trichinella). She deposits her batch of offspring inside the blister she induces at the level of her head and birth canal. Blister formation takes place over several weeks, after which time the host begins to experience a sharp pain in that general area. Relief is found by submerging the affected area of the body, along with the blister, into freshwater. This is just what the worm wants its host to do. So a trip to the well serves two purposes for the host and one for the parasite. Once immersed in water, within minutes the blister rapidly swells then bursts open, releasing its complement of larvae. This reduces pressure on the host tissues surrounding the blister and temporarily relieves the pain and suffering of the victim.

The next step in its life depends on these step-wells maintaining a small assemblage of aquatic life (each one is a miniature aquatic ecosystem). Like oases, step-wells can be a source of malaria since in most of them mosquitoes can breed continuously throughout the year. In addition, the well water supports other life, including microscopic crustaceans, just like the ones that serve as intermediate hosts for the fish tapeworm (see chap. 6). As the roundworm larvae begin to swim, copepods in their general vicinity see them wriggling and ingest them, thinking of course that these slender life forms are just another food item. Bad move! Just like the crustacean that swallowed the coracidium of *D. latum*, this copepod is now also victimized, being forced to support the life of the worm until it encounters another human host. This last step may seem like a long shot since many step-wells are large, and the chances of including water with the copepods and the larvae into a water-carrying vessel are small. But recall that our sufferer was seeking relief from the discomfort of the infection *and* water for the village. Scooping up a large bucket of water next to one's feet while they're cooling under the water's surface seals the deal for the host and the parasite. It may take several days to return home; all the while the parasite is completing the last phase of its development inside the copepod. Once the drinking water is finally available to those in the settlement, everyone who drinks from that container has an opportunity to become the next host for *Dracunculus medinensis*.

After the infected copepod is swallowed, the infective larva is digested away from the crustacean in the stomach and is carried to the small intestine where it penetrates the tissues and migrates within the connective tissues. *D. medinensis* continues to grow and develop in that region of the body, taking up to a year to do so. During that time, the worm sheds its skin (cuticle) several times, becoming either an adult male or female. The worms mate, and the now fertilized female, filled with live larvae, crawls down to the lower extremities, where it induces a blister near its head. The life cycle is now complete. We know nothing about the details of the worm's life during its developmental phase once the infected copepods are swallowed. For example, what happens to the males after they mate? In fact, where do they mate? One big unanswered question is why we tolerate the infection so well. This large worm must somehow control our immune system in a way that allows it to survive for up to a year and carry out its biological imperative, reproduction.

Let us now return to the subject of control programs for Guinea worm disease. More specifically, I want to describe the events that led up to an issue that was dead-on for a familiar curse that states "Be careful what you wish for, lest it come true," and a saying that often results in negative unintended consequences: "What they don't know won't hurt them." In 1981 the World Health Organization initiated a global effort to greatly reduce diarrheal diseases in rural populations. It prepared a statement that read:

INTERNATIONAL DECADE FOR
CLEAN DRINKING WATER, 1981–1990

In the 1980s, there were 1.8 thousand million people living in the rural areas of developing countries. Only one person in five had access to clean water. 590 million (41 percent) of the children under 15 years old did not have clean water. In the developing countries, one hospital patient in four suffered from an illness caused by polluted water.

Daily, millions of women and children have the chore of fetching water, taking up to half a day and using energy which would have been better spent on education, training, or simple survival. Even then, the water may not be clean, yet contaminated water means sickness and death.

WHO had always known that step-wells were a major point source for diarrheal diseases and was also convinced that these stagnant bodies of water were the most common sources for Guinea worm infection. If the organization could persuade users of the wells to close them over and convert to using hand pumps that WHO agreed to install next to them, both problems would be addressed. It should be stated here that WHO considered diarrheal diseases, and the infant mortality they caused, a far more serious issue facing the world at large than Guinea worm. But this ambitious program encountered a major snag shortly after it began. It involved withholding vital information as to the real reason for covering up the wells. The health authorities told the users of those wells only that they wanted to control Guinea worm disease. Many people who routinely used the drinking water sources expressed a deep sense of anxiety over the prospect of having their social gathering places eliminated. Information as to what happened next is a bit fuzzy, but the end result was not. They began to inquire about how *D. medinensis* was transmitted. Some even wanted to know the details of its life cycle. This would prove the undoing of the goal of the original program—namely, the reduction of infant mortality from drinking contaminated water. When the part about the integral involvement of small crustaceans in the life cycle of the worm was described, a few of those who vehemently opposed the closure of their coveted step-wells conceived a clever solution, which proved 100 percent effective in removing the crustaceans, and without the need for closing down the step-wells.

Thin, gauzelike fabric is used to make traditional clothing in many regions in India. It is cheap to manufacture, and most Indian women wear garments made from it. Early on in the control program, it was discovered that if this readily available material is folded over on itself to create four layers, the crustaceans found in well water can be easily trapped in the cloth by passing the water through the "filter" into another clean bucket. The life cycle was interrupted at that level. Game over! Villagers could keep their step-wells and get rid of Guinea worm, too. Unfortunately the microbes that typically cause diarrheal diseases passed right on through. The result was a dramatic reduction of Guinea worm disease without any significant reduction in the infant mortality rate. Word of the prevention method spread rapidly and became the gold standard for Guinea worm prevention. Today the control program advocated for by the Carter Center, founded and led by former President Jimmy Carter, involves the use of a special cloth filter made available free of charge to whoever needs them. The Centers for Disease Control and Prevention lists on its web site (as of 2012) the following method most commonly used to prevent infection with Guinea worm: "Often there are no safe water supplies in these villages. Therefore, fine-mesh cloth filters are given to households to strain out the infected copepods (tiny "water fleas") from contaminated drinking water. Copepods live in stagnant water and become infected when they eat Guinea worm larvae released into the water."

In summary, *Dracunculus medinensis* may soon become an extinct species. On the other hand, diarrheal diseases caused by a wide variety of microbes, commonly found in polluted drinking water supplies throughout the tropics, will continue to take their deadly toll on newborns and infants. One possible alternative that might have salvaged the program could have been to condemn the step-wells for drinking water and install hand pumps next to them, but leave the wells uncovered. They could then still function as social networking places, and drinking water could be obtained from a safe-to-use, hand-pumped bore hole well strategically placed a convenient distance away from them. I am certain that this approach was suggested many times and rejected as being too complicated. Still, telling a consumer public the whole truth about any specific program designed to control a broad spectrum of infectious agents would have enabled that health agency to enlist the trust and support of local leaders, thus facilitating the implementation of complex solutions to complex problems. It is apparent from this story of dracunculus control that a more sophisticated and inclusionary philosophy must be built into the design of all future community-based health interventions, and in many instances this has been done. People—even those who have not had the advantage of formal schooling—are a lot smarter than most governmental agencies give them credit for and can solve complex problems if given all the facts. So be very careful what you wish for! In the end, as William Shakespeare remarked, "All's well that ends well." In the case of Guinea worm control, it might have been better to say "All's well that ends wells"!

Plowshare Concepts

How does *D. medinensis* avoid inmune attack and live for up to a year inside us without any apparent ill effects on its biology? This question is obviously not a new one to the reader, as many of the parasites already discussed also have learned this trick. Regardless of this common feature, however, it is apparent from numerous laboratory-based studies that each species has evolved a unique strategy to do so. The genome of dracunculus has not yet been completely revealed, but if and when it is, we will then be able to search its sequence database for parasite proteins with known interactions with our immune system (e.g., regulation of interferon production, IgE synthesis, eosinophilia and homing responses, antimitosis, modulation of the Th1 and Th2 systems). Cloning putative *D. medinensis* immune function–altering genes into related filarial worms that have much shorter life cycles, making them convenient to work with in the laboratory (e.g., jirds and *Brugia malayi*), might give important clues as to the escape mechanisms used by this parasitic worm. These genes might also give us insight into human immune disorders that could be controlled using products derived from a nearly extinct parasite. We may well proceed to kill off the last remaining *Dracunculus medinensis* parasites through control programs already in place throughout the world, and this is all to the good. But let us keep its cDNA library in the lab deep freeze and at the ready, so this dreaded infection can start giving back to us some of the benefits it enjoyed throughout the millennia of its long presence on Earth.

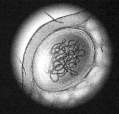

8   NATURE HAS ALL THE ANSWERS. WHAT'S YOUR QUESTION?

What follows are my last two parasite stories, and that is just what they are: stories. I have put them together this way because they deal with various aspects of drug discovery, and they also have application to the field of medical parasitology. I wrote each one as a historically accurate fictional accounting of how these two parasites carry out their lives, the human lives they affect, and how they have contributed in major ways (albeit differently) to the development of therapeutics that are currently in use. While the characters in each episode are fictitious, the places in which their lives unfold are real, so despite all the progress we seem to have made as a species over the last one hundred years of our evolution, events similar to what I am about to describe can unfortunately still be observed throughout those regions of the developing world.

Luck Favors the Prepared Mind: *Onchocerca volvulus*

The morning had arrived all too soon, hot and humid, not unlike all the other days in this broken-terrain part of equatorial West Africa. For one small boy, today was indistinguishable from all the others he had experienced over the past six years of his young life. He was the most important person of a daily parade consisting of himself and five weathered old men, all hanging on by one hand to a long, thin stick, worn shiny-smooth through years of use. The boy held on to it with all his might. The child could not have been more than twelve, yet he was already an acknowledged leader in his tiny village, located some five miles east of Gummi in the northernmost upcountry region of Nigeria.

The troupe of six marched as one in slow cadence, single file, at a pace that could have easily led the party to be mistaken for a funeral procession. No one spoke. This sad row of tired, expressionless faces and listless bodies was, in fact, a quiet celebration of the indomitable spirit of life at its most basic level of expression: they were all going to work. Did I mention that everyone in the troupe was blind? Everyone, that is, except the boy.

These proud, resolute men and their guide went away from their homes and families every day to work the fields several hundred yards away on the outskirts of their village. Their town was not nearly as densely populated as it had been in the past. Many had long since left, and those who could not remained on to bear the ravages of a dreaded disease that literally flew out of the river and sought them out, haunting them mercilessly to their graves. The same small, swift-flowing stream that was the source of all their misery was, ironically, also their sole source of life-sustaining freshwater.

When they arrived at their final destination, the adolescent led the men to a work-station delineated by a waist-high rope, strung taut from one end of the field to the other. It served as a guide for the laborers. The boy positioned the work force evenly, about 2 feet apart, as one might do when accommodating certain plants that demanded space in which to grow. He then handed each worker either a wooden hoe or a rake that he selected from a pile of tools left at the work site. In prior years many more workers had been available, but owing to the steady exodus of residents, high infant mortality from devastating outbreaks of malaria, and the occasional horrific epidemic of Lassa fever (a viral infection with a 30 percent mortality rate, acquired by exposure to rat urine), just these few adults remained to carry on.

One by one, the boy assigned each man to his daily routine of digging or raking dirt over the hole made by the worker next to him. Each held a tool in one hand and the rope in the other. At the boy's command—two claps of his small, calloused hands—work began. With a single, forceful downward stroke, one man gouged out a hole, then the boy, crouching crablike, followed with a seed, usually corn or millet, that he selected out of his tattered, shopworn pants pocket. Still in position, he tapped lightly on the foot of the man holding a rake next to the hole-digger. In two swift moves, the wooden rake's tines covered up the seed. Then the boy stood up and stomped on the same spot as hard as he could, raising a cloud of dust in the process, and at the same time compacting the loose soil in hopes of making it more difficult for birds and small rodents to raid the hard-earned plantings after the workers left for home, just before sundown. Despite their best efforts, even in good years, less than half of what was planted survived to harvest.

Work progressed steadily, but at an agonizingly slow pace. When the first row was finally sown, the men took a small break, as the boy moved the rope to the next set of posts and, after realigning the work force accordingly, brought his hands together twice more with the same result. This was repeated many times over during the day, with a brief moment for lunch to refuel, rest, and break the monotony. Irrigation was an absolute necessity. Water came from the same stream near town that also ran adjacent to the last planted row of the farm. Tragically, the proximity of running water to their workplace ensured the loss of their sight. That was because all day long the workers, including the boy, suffered numerous bites from black flies (fig. 8.1) that bred in their precious stream. The flies carried a dreaded curse: *Onchocerca volvulus*. This was the root cause of their suffering.

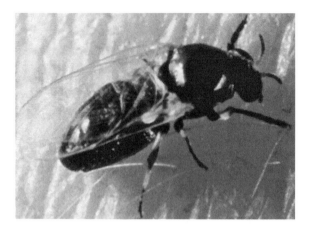

Figure 8.1 The black fly, the vector of *Onchocerca volvulus*

Blindness was inescapable. Every man over the age of twenty was completely without sight. All hope was gone for any of them to live a normal life. The disease was known to those unfortunate enough to experience it firsthand as river blindness.

The scenario above was common throughout West Africa before the European explorers arrived, and it persisted right through to the 1970s. Then a miracle happened. A silver bullet was discovered that changed forever the fate of all those subjugated under the heavy burden of onchocerciasis. A new drug, ivermectin, became the weapon every-one had been looking for to drive out this evil presence once and for all. Before that day of reckoning, controlling the population of black flies with DDT was the only reasonable approach to reducing the inci-dence of onchocerca infection. It involved the use of insecticide in all running waters harboring the filter-feeding black fly larvae. But, along with the black flies, this potent organic chemical also killed off all the other stream insects, short-circuiting the ecological associations of all the other life forms living there, especially the fish that fed on these insects. Needless to say, the local fisheries suffered greatly during this period of intervention (1960–1980). Nevertheless, the generous use of DDT, applied both manually and by airplane, was an effective method of reducing the mortality and morbidity from this worm infection. Programs in every affected West African country using this approach were heavily supported by the World Health Organization and other, less well-known nongovernmental agencies. At one point during the 1960s, civil war was raging in Nigeria, but that did not stop the WHO's aerial spraying programs. Not one shot was fired by either side at these low-flying spray planes. That's how important everyone felt the attempts at controlling this awful disease were.

Black flies are aquatic insects that breed in cold, fast-running fresh-water rivers and streams. In addition to Africa, these pesky biting flies are found in many other places throughout the world, including North, Central, and South America. Anyone who has spent time in the forests of New England in May and June knows all too well the persistent biting habits of black flies. In the New World, the geographic distribution of onchocerca matches the distribution patterns of slavery, with one major exception: the United States. There are few environments in the southeastern part of the United States that meet this ecological requirement for cold running water where slaves also lived and worked. At one point river blindness affected the lives of millions of people living in the rural areas of West Africa and parts of Central and South America. Altogether, nearly 120 million people lived within the flight range of the black fly vector.

Note that I have intentionally used the past tense when referring to the status of its worldwide prevalence. In fact all that has changed, and decidedly for the better, within just the past twenty years. How that came about is the central theme of this story.

First, though, let's learn a bit more about the worm and the diseases it causes (fig. 8.2). The adult parasites are long, slender worms—females

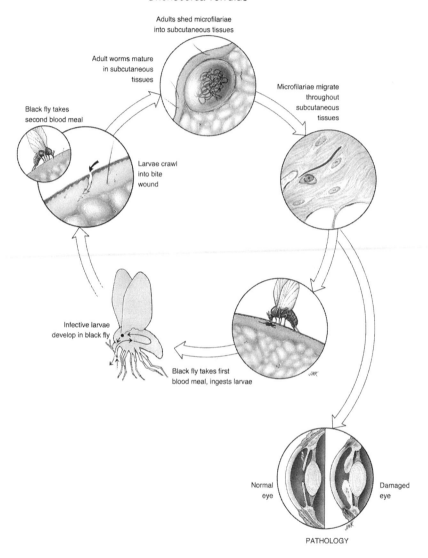

Figure 8.2 Life cycle of *O. volvulus*. Illustration by John Karapelou. (*Source: Parasitic Diseases*, 5th ed., (c) Apple Tree Productions, LLC, New York)

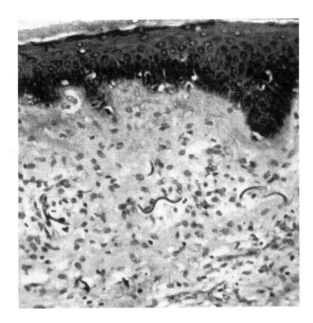

Figure 8.3 *O. volvulus* underneath the skin

are 30–35 centimeters in length; males are 40 centimeters long—that live coiled up in a ball inside the subcutaneous tissues, just underneath the skin (fig. 8.3). They induce the host to produce a nodule of tissue, often reaching the size of a walnut. It has the appearance of a tumor, hence the name *onchocerca*. Inside the nodule there can be as many as fifty adult worms, but the usual number is between three and five, and the nodule's size varies with that number; the fewer the worms, the smaller the lump under the skin. If the nodule is directly over a bony process, like the point of one's hip or elbow, or anywhere on the head where the skin is thin, then it can easily be seen. Otherwise *onchocercomas*, as they are referred to by physicians and health workers, can be located by thoroughly palpating the body and feeling for unusual lumps of tissue. The adult parasites in them induce the host to supply them with an extension of the circulation, similar in many respects to the process that trichinella larvae engineer during the formation of the Nurse cell (see chap. 1). In this way *O. volvulus* can obtain its nutrients and get rid of its metabolic wastes without moving an inch. They are the ultimate homebodies of the parasite world and can survive in their nodular homes for up to ten years without being disturbed. There is little information regarding the parasite's day-to-day needs, so just what it is in the way of metabolites (amino acids, glucose, vitamins, fats, etc.) they derive from us and how they do it remains to be revealed.

With heavy infections, a person might harbor over one hundred of these nodules, which may be distributed all over the body—wherever the black flies have access to exposed skin to obtain their blood meals. In Central and South America, the nodules tend to be located above the neck, since long-sleeved tops and pants are more commonly worn there than in Africa, where it is customary to wear much less in most rural settings. Nodules in the groin can block the flow of lymph, as in lymphatic filariasis (see chap. 4). A condition known as "hanging groin" can result, which is nearly indistinguishable from the most severe form of *Wuchereria bancrofti*, the causative agent of elephantiasis.

If the adults were the only parasitic stage to live in our tissues, then we would have only these medical problems to deal with. But that is obviously not the case. So if it's not the adult stage that causes the most serious illness—blindness—then what is responsible for all the damage leading to the loss of sight? The answer is the offspring, the microfilariae (fig. 8.4). Adult female worms produce about one thousand of this larval stage each day for up to ten years. That means that every female worm is the proud mother of 3,650,000 babies! These immature onchocerca are the real culprits, despite their diminutive size. But, as is true for all things in nature, deciphering the molecular details underlying the mechanisms responsible for the pathological effects of the microfilarial stage on the host was not straightforward, nor in the end was the explanation simple.

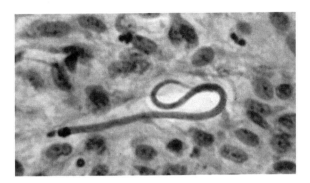

Figure 8.4 *O. volvulus* offspring, the microfilariae

Microfilariae are not stay-at-home kids. They are born with a wanderlust. Each one has the option to travel anywhere within the subcutaneous tissues and experience the host up close and personal. Since this tissue is anatomically contiguous with many other parts, including the eye, it's no wonder that the parasites are such troublemakers at this stage. We inadvertently help out, too, by doing what we can to get rid of them. Our immune system, as good and useful as it is, tends to overreact to unwanted intruders, damaging normal tissues as well as neutralizing the offending substance or microbe. We call this the "innocent bystander effect." But there is another coconspirator that we must take into account: *Wolbachia pipientis*, a bacteria that lives only inside the cells of the adult worm and its offspring (see chap. 4). Astonishingly, this much smaller bacterial life form significantly aids and abets the worm's penchant for stirring up mischief and mayhem at the cellular level.

The microfilariae start their odyssey, setting off in all directions away from the home nodule, and their trip is not without purpose, either. They are hoping to become ingested by a black fly and continue the life cycle. When a black fly bites, it sucks up blood, and in a person with a long-term onchocerca infection, the insect inadvertently also ingests some microfilariae. Both male and female black flies feed on blood, unlike their mosquito relatives, in which only the female needs a blood meal for egg production. That means that every black fly can transmit the disease. Once inside the stomach of the black fly, the larva first penetrates into the fluid-filled space between the organs (hemocoel) and then penetrates into the flight wing muscles. There, the parasite develops into the infective larva for another human. This takes about five to six days. It then makes its way out of the muscles and into the mouthparts of the insect, where it waits until the infected fly takes another blood meal. As the black fly feeds again, it moves its bladelike mouthparts back and forth, literally tearing a hole in the skin. This is painful, as anyone who has suffered bites from these aggressive insects knows. When the blood begins flowing, the body heat from the host stimulates the larvae in the fly to migrate out through the mouthparts and onto the skin. As the fly leaves the site of the bite, the larvae crawl into the wound and establish a new homestead in the subcutaneous tissues. Eventually a nodule forms in response to the growing worms. Soon thereafter, the worms mate and the females begin to shed microfilariae. The life cycle is now complete.

Meanwhile, back in the host, our immune system begins to sense trouble in the form of wolbachia metabolic by-products (antigens) and attempts to kill off both the microfilariae and their hidden cargo. Our white cells join the rumble. Regrettably, all this immune activity has little effect against the infection. Over time (months to years), the entire outer skin becomes leathery and pigmented, and an intense and permanent itching sensation is established over the entire surface of the nodule. Imagine trying to get in a good night's sleep with all that going on! Over a period of several years, the eyes eventually become involved. When these immune reactions occur inside the eye, all the parts of that sensitive organ begin to fail. The lens clouds up with immune complexes (combinations of antibodies and foreign substances). The muscles controlling the shape of the lens that helps us focus begin to weaken. The retina becomes damaged by our own immune cells, trying their best to kill off the larvae and their evil little symbionts, but to no avail. All this pathological activity eventually leads to total blindness.

We have only recently become aware of the significance of this odd couple—that very strange bacterium and the microfilaria—even though these symbionts were first observed together under the electron microscope in 1977 by a friend and colleague of mine, Dr. William Kozek. In fact, it seems like every time we look closer at any host-parasite relationship, the more we come to realize how evolved and well adapted parasites are in their niches within us, and how complex their lives are. As I will describe later on, killing off just the bacteria with antibiotics has serious consequences for the worm and the host and is one course of treatment that's currently used, but the most common treatment is the miracle drug ivermectin.

Before the advent of ivermectin, besides those insecticide-based control programs that targeted only the vector, other public health initiatives went straight after the parasite itself, even though no effective drugs were available. Treatments were based on the efficacy of periodically removing the nodules containing the adult worms. For instance, in the mountainous countryside of affected areas of Central America (e.g., Honduras, Costa Rica, Guatemala, Panama), teams of health professionals would roam about visiting coffee plantations and remove en masse as many nodules from the heads of infected agricultural workers as they could in a day. They applied sterile technique and local anesthetics to ensure that the patients were comfortable throughout the nodulectomy procedure. Many teams even entered into good-natured contests to see who could collect the most nodules in a given amount of time. This activity had a measurable effect on reducing the amount of blindness among these populations, since, as you may recall, the closer the nodules are to the eye, the more likely it is that blindness will ensue. But episodes of unforeseen human conflict frequently conspired to thwart even those best of intentions. Nodulectomy teams were periodically disrupted from carrying out their charge by rashes of armed conflicts, bouts of civil unrest, and shifting territorial battles among warring illicit drug-trafficking cartels. All that changed for the better in the 1980s.

Let's take a moment, before starting our discussion of ivermectin, for a brief look at how drugs, particularly antibiotics, were and still are being discovered, and how they become developed to the point of commercial success.

Traditional medicine chests (see below) are filled to the brim with curatives derived directly from nature. Throughout human history, many plants and animals have been tested for their effectiveness in treating just about every known aliment, both for humans and for their domestic animals. But even the remedies that worked were never purified and characterized. The burden of periodically going out and harvesting these often rare life forms from their natural habitats made them even scarcer and, in some instances, used them up altogether. Then, in 1899, the Bayer Company in Germany, a manufacturer of dyes and stains, announced that it had successfully made aspirin (acetylsalicylic acid) in its laboratory, a compound originally isolated from the willow tree. This first synthetic drug established the pharmaceutical industry, and it has never looked back. It was much later, however, when the first antibiotic was identified, purified, and used to effectively treat a disease. That was in 1939, when René Dubos, while working at the Rockefeller Institute for Medical Research, stated his belief "in principle of 'antibiosis' and the supremely simple working hypothesis that soil as a self-purifying environment could supply an agent to destroy disease-causing bacteria." Dubos, a soil ecologist by training, eventually succeeded in growing a soil-dwelling microbe that secreted a substance that proved lethal for almost all other known microbes, including those that caused diseases in humans. He called the substance gramicidin. Unfortunately this compound killed many varieties of normal cells, too, including ours, reducing its usefulness in treating infectious diseases of the deep tissues. It remains in use today as a topical ointment for superficial wounds.

The first effective, safe-to-use antibiotic was penicillin, discovered entirely by accident by Alexander Fleming. One fateful day in 1928 while working in his laboratory, he noted that a certain greenish mold had inadvertently contaminated some of his petri dishes, on which were growing yellow-colored colonies of the bacteria *Staphylococcus aureus* (*aureus* means golden), a microbe known to cause numerous clinical illnesses, some of them fatal. He noticed that the bacterial colonies nearest the mold on the plate were lysed and dead, while those farthest from the mold continued to grow. He recorded his results, describing how the mold must have killed off a part of the bacterial culture by secreting something soluble onto the plate, since he could see a clear zone next to the mold that kept the bacteria from growing up next to it. We now refer to this as the zone of inhibition. He continued to test out his new-found mold on many other types of bacteria and found that some of them were killed, while others remained unaffected. He published these observations in 1929 and then moved on, choosing instead to address other scientific problems related to the bacteria, not the mold that had done them in. The world quickly forgot about Fleming's discovery until Dubos published his results much later on. World War II was imminent. Dr. Howard Florey and colleagues in England became rejuvenated by Dubos's findings and rediscovered Fleming's work on penicillin. In Fleming's own words: "It was only when some 10 years later after the introduction of sulphonamide had completely changed the medical mind in regard to chemotherapy of bacterial infections, and after Dubos had shown that a powerful antibacterial agent, gramicidin, was produced by certain bacteria, that my co-participators in this Nobel Award, Dr. Chain and Sir Howard Florey, took up the investigation."

The rest, as we are so fond of saying, is history. I think that is why Louis Pasteur's famous quote is so apt to mention in these situations: "Luck favors the prepared mind." Penicillin's incredible effectiveness proved out over the later stages of World War II in use among wounded Allied troops and civilians. Before penicillin, Western medicine relied on sulfur compounds and the maggots of certain flies for the treatment and debridement of open wounds. These crude medical strategies date back to the 1800s and the Civil War. Some were effective if applied early enough after the wound was incurred. However, most injuries had a time delay of several days before receiving proper attention, owing to the usual absence of medical personnel at the battlefront. By then gangrene (a common bacterial infection of deep wounds) had usually set in, eventually either killing the patient or resulting in an amputation, if the patient was lucky. For a good read on the subject written in graphic, eloquent prose, pick up a copy of Stephen Crane's *Red Badge of Courage*. The effectiveness of penicillin in World War II instantly made it a household word and convinced even the most skeptical of physicians that it was here to stay.

Several Nobel Prizes were awarded for the discovery and application of penicillin to a wide variety of infectious diseases. But even then, some of the possible limits of the drug were known. Again, quoting from Fleming's Nobel lecture: "But I would like to sound one note of warning. Penicillin is to all intents and purposes non-poisonous so there is no need to worry about giving an overdose and poisoning the patient. There may be a danger, though, in under-dosage. It is not difficult to make microbes resistant to penicillin in the laboratory by exposing them to concentrations not sufficient to kill them, and the same thing has occasionally happened in the body." How prophetic were his words of warning, as over the next thirty years, more and more pathogenic bacterial entities achieved resistance to this once-lethal drug.

In the meantime, spurred on by the tremendous success of penicillin, a new era of drug discovery was in full swing. Streptomycin was developed at Rutgers University in 1943 by Selman Waxman, the same person who had mentored René Dubos (small world). Waxman was the first to use the term "antibiotic" to describe the activity of his new class of compounds. After the end of World War II, New Jersey became the epicenter of the pharmaceutical industry, hosting some of the largest companies in the world. As of 2012 there were no fewer than seventy-nine drug-manufacturing companies listed on the Internet that called the Garden State home. Among the more notable ones were Bristol-Meyers Squibb, Hoffman-La Roche, Merck & Company, Novartis Pharmaceuticals Corporation, Schering-Plough Corporation, and Warner-Chilcott Laboratories. Profits soared with each new discovery of a fermentation product derived from a novel soil microbe—and many quickly made it to the marketplace and were touted as the next "miracle" drug by sales reps of the pharmaceutical industry and physicians who used (and sometimes abused) these new drugs until they became ineffective. "No problem," the industry said, "we'll just isolate another one in a couple of days and things will be right as rain again, just you wait and see." Today there are a total of 138 unique antibiotics for sale throughout the world. Most of them no longer work as well as they did the moment they were made available. Is this a problem? Yes, indeed, it is a big-time problem.

This is the darker side of the pharmaceutical industry that was the specter of Fleming's ominous prediction. Biology is biology, no matter what the life form, no matter what the situation. Nature has all the answers. What's your question? René Dubos's question was simple. Can I isolate a compound from a soil-dwelling microbe whose purpose is to inhibit the growth of other microbes? Of course, we now know that the answer was a resounding "Yes!" All he had to do was devise a way (called an assay) to find it. His solution was to grow soil microbes in a liquid growth medium in flasks (so-called fermentation flasks). He then looked for active compounds in the liquid part of the culture by testing concentrates of the liquid on cultures of bacteria growing on petri dishes. Every time he got inhibition of growth on the plates, he purified the compound from the liquid culture, and in this way Dubos eventually discovered gramicidin.

r every answer, there is yet another question to be asked. The microbe whose ox just got gored is allowed to ask a question, too. "Can I acquire (by natural selection) a way around this antibiotic activity so I can continue to grow and out-compete all the other soil microbes?" The answer here is, again, a resounding "Yes!" The mechanism is called *antibiotic resistance* and is just one more example of natural selection at work. In nature, this arms race has produced all of what we see before us as life on Earth. Every creature has its role to play, and none has a permanent advantage over any of the other life forms, owing to the constant rate of mutation of the DNA molecule. This results in biological diversity and an ever-changing physical environment, which selects for those mutants best fit for that particular environment. With regard to microbial life, mutations within a colony of millions of individual bacteria that are susceptible to a particular antibiotic, let's say penicillin, occasionally result in a few that are now resistant to that drug (see chapter 3 and the life of African trypanosomes for a similar story). The ones that cannot live in the presence of the drug then die, making way for those few that can. The next thing you know, the entire plate is again covered with bacteria that can grow right up to the green mold on the same plate. That is exactly what Fleming observed, and that is exactly what occurred on a global scale when we began to use these antibiotics indiscriminately in lower than therapeutic doses, selling them over the counter to anyone who wanted them. No medical prescription needed.

The result? Today there are just a handful of infectious agents that are still fully susceptible to most antibiotics, and a few are now resistant to all of them. Tuberculosis is one of the most dangerous organisms on the planet. In India and Russia, as of 2012, there were known strains of TB that did not respond to any known chemotherapeutic agent. Unless the global pharmaceutical community comes up with a new set of products, and soon, millions of people, mostly the poor, will die from this infection. Other resistant organisms that come to mind include malaria, gonorrhea, and many types of staphylococcus infections.

Ivermectin is currently the gold standard for treating onchocerca infections. It works by binding with high affinity to glutamate-gated chloride channels that occur in invertebrate nerve and muscle cells, causing an increase in the permeability of the cell membrane to chloride ions with hyperpolarization of the nerve or muscle cell, resulting in paralysis and death of the microfilariae. It has no effect on the adult worm, so a single annual dose of the drug must be given until the adult worms in the nodules die of old age, which can be as long as eight to ten years. Regardless of this obvious drawback, over the past thirty years, ivermectin has been directly responsible for the complete elimination of river blindness in most of equatorial Africa, and in many regions in Central and South America as well. That is because there are no other animals besides humans that can harbor the infection (i.e., no reservoirs). So how did we come to discover this drug, and how was it eventually made available to all who needed it, and free of charge? It's quite a story.

To begin with, drug development requires an enormous amount of effort and time, to say nothing of the actual expense. Most new drugs come from the research laboratories of the major pharmaceutical houses such as Pfizer, Merck, and GalaxoSmithKline. It currently takes many years to bring a promising compound from the bench to the pharmacy, even after all the science has been worked out and scale-up production methods have been established. That is because between 1957 and 1961, the drug thalidomide, a sedative used extensively by women during various periods of their pregnancies to alleviate morning sickness, induced teratogenic effects on fetal development, resulting in the loss of functional arms, and in some extreme cases the loss of all four limbs, in newborns. All this happened after the drug had been tested extensively on numerous animals, including nonhuman primates, without encountering any of these tragic side effects. Apparently only humans are susceptible to them. The entire process of drug development and marketing was forever changed. Clinical trials became more rigorous, and the amount of research needed to assure the Food and Drug Administration of a drug's safety for use in humans increased exponentially.

Ivermectin was discovered in May 1975. It was an active component of a fermentation broth made by a soil organism present in a sample obtained by Merck Sharpe and Dome Research Laboratories in Rahway, New Jersey, from the Kitasato Institute in Tokyo. These two companies had previously agreed to collaborate in the discovery of new antimicrobial agents, employing essentially the same kind of technologies as those pioneered by René Dubos and Selman Waxman. The reasoning was that if it worked for them, then it should work for us, too. Lo and behold, in this case, luck favored the prepared company, and ivermectin was soon on its way to becoming the newest silver bullet in the fight against river blindness. But there were many steps along the way to moving ivermectin from the discovery phase to the marketplace. It all hinged on the assay.

In this instance, Merck had decided to set up a multipronged approach to analyzing all fermentation products, which included a tandem screen for antiparasitic agents. For this assay, established in the Merck laboratories in Rahway by Dr. J. R. Egerton under the direction of Drs. William Campbell and Ashton Cuckler, an intestinal nematode parasite of mice, *Nematospiroides dubious*, and a protozoan parasite of chickens, *Eimeria tenella*, were used. When the liquid phase from a fermentation flask containing the Japanese sample OS3153 was mixed in dilute amounts into the drinking water of mice infected with *N. dubious*, the investigators found one mouse that did not have any infection, and the others in the same cage had reduced worm burdens. This was the result they were hoping for. After repeating the experiment with higher numbers of animals and various levels of the broth, a solid set of data emerged confirming the presence of a new compound capable of reducing the presence of nematode parasites in experimentally infected mice. After many years of work following that seminal finding, ivermectin, a complex chemical with a broad spectrum of activity against invertebrates whose biochemistry differed significantly from mammals, had been discovered, purified, and produced in mass quantities. Its safety was proven for use at high doses for domestic animals of commercial value, and clinical trials proved that it was also safe for use in humans.

In looking back, the association of Merck with the Kitasato Institute was key to its discovery. The records indicate that Japanese soil sample OS3153 had been collected in close proximity to a nearby golf course, as a part of a global effort to cast a wide net around the natural world, enabling subsequent investigators the luxury of becoming lucky. Ivermectin might never have been developed if someone had not committed to this seemingly random approach to drug discovery, yet the history of antimicrobial therapies is filled with such stories.

What happened next is a rare example of altruism at its best. Ivermectin was eventually shown to eliminate microfilarial production by *Onchocerca cervicalis*, a parasite of horses similar in biology to *O. volvulus* in humans. Small clinical trials in patients suffering from river blindness followed shortly thereafter. In West Africa ivermectin proved to be well tolerated and reduced the production of microfilariae to zero. However, it was quickly realized that no developing country could afford to purchase it. That is when Senator Bill Bradley of New Jersey, former professional athlete, brilliant politician, and humanitarian, petitioned Merck to give the drug away to those people who needed it the most. Sales in animal health–related areas had by the mid-1980s made hundreds of millions of dollars in revenue for Merck, so it could easily afford to do so, but up to then, no drug company had ever even contemplated literally giving away one of its prized compounds. Dr. Roy Vagelos, CEO of Merck, agreed with Senator Bradley, and together in 1987 they made history, with Merck becoming the first pharmaceutical entity to give away a drug on a global scale for compassionate use. Since then other companies have followed suit, setting a new level of humanitarian giving for the rest of the industry to follow. The only caveat was that the countries accepting the gift of ivermectin (now branded as Mectizan) had to keep track of the prevalence and incidence of river blindness and make the data available to both the Merck Drug Donation Program and the World Health Organization. As of 2012 nearly 700 million doses of the drug have been given, and each year some 80 million people receive it free of charge. The results have been absolutely phenomenal. Very few endemic areas remain in Africa and Central and South America. With this kind of pressure on transmission, and no detectable drug resistance in the parasite to impede the use of ivermectin, it is entirely possible that within a few more years river blindness will become the third infectious agent ever to be made extinct on purpose (smallpox virus being the first and rinderpest virus the second).

There was one other advancement that made those suffering from this dreaded disease stand up and cheer. By treating patients harboring active onchocerca nodules with doxycycline, an antibiotic designed to kill off the wolbachia bacteria, but without giving any other therapeutic agent that would directly affect the survival of any stage of the worm, voilà, the females ceased the production of offspring and eventually died, as if they had been treated with ivermectin. That finding alone should keep researchers happy for a long time trying to find out why that should be. This result proved beyond anyone's doubt that all the pathology from chronic river blindness was due to our immune system malfunctioning! It was actually reacting to the by-products of the parasites' wolbachia symbionts, not against the worms themselves. What a surprise! Other antibiotics were tried, too, but with limited success. These findings are counterintuitive, based on the traditional view of the hosts and their parasites. I think what all this really suggests is that we still have a lot to learn about life on Earth, particularly when it comes to host-parasite relationships.

Traditional Drugs and Their Side Effects:
Spirometra mansonoides and sparganosis

The perfume-laden air of dusk clung to the rolling landscape like a warm, moist blanket. It was midsummer on the Chinese island province of Hainan, and the second crop of the year stood tall and verdant in the still flooded rice fields. On the outskirts of the tiny village of Qianjia a lone farmer stood on the grass-covered berm next to a rice paddy in the growing darkness. Earlier that day, Lian-Fu had worked hard in these same fields, as he had done for the past twenty-three years. But the rice farmer did not return that evening to continue laboring in his assigned plots. Rather, he was on a mission. His sole aim was to save his young son Gao's left eye from going blind. To do so, he must first catch a frog whose healing powers were known to everyone who lived in that remote region.

A week earlier Gao had accidentally fallen on a sharp stick while playing with his classmates outdoors during recess. It had pierced through the upper lid of his left eye, narrowly missing the globe. The puncture became infected, causing him much pain and anguish. On the day the accident happened, his teacher had placed a handkerchief over the wound to control the bleeding and sent the child home, for there was no hospital or trained physician or even a nurse in their tiny hamlet to take over the care of his injury. Neither were there any pharmacies nor prescription drugs available. Over the next three days, Gao's mother, Su, tried everything she had been taught by her own mother in dealing with situations such as this, but nothing she did eased the child's pain or stopped the spread of the bacterial lesion that was taking over Gao's eye. The infection worsened with each passing day. Finally, realizing that her efforts were in vain, Su urged her husband to seek advice from the local healer. After briefly explaining the problem to their shaman, the prescription came as no surprise: a poultice made of freshly prepared frog tissue. The old man emphasized to Lian-Fu that using the right kind of frog was essential, and in careful detail he described the ways to identify it, down to its characteristic bulging eyes and mottled olive body markings. Its scientific name is the tiger frog (Rana *rugulosa*).

Three previous nights of unsuccessful hunting had resulted in nothing more than an increased level of anxiety and sleepless nights of worrying over his son's now acute illness, punctuated by the child's continuous, pathetic, low-pitched moaning. Hoping tonight would be more fruitful, the solitary hunter continued his progress along the raised bank of the paddy in slow motion, heronlike, crouched and ready to spring into action should he catch sight of the medicinal amphibian. One clumsy step might send all the frogs within earshot to seek cover, deep into the soft muddy bottom of the man-made pond. Then there were the ever-present distractions to worry about. In addition to chasing down and capturing what was proving to be a highly elusive prey, he was also being very careful not to surprise any of the other predators in the area that might be dining alfresco—cats, waterfowl, and such. While these harmless animals were his main competition for the evening, a few others could prove fatal. Take Russell's viper, for example. That deadly snake was a regular inhabitant of the rice fields, feasting on frogs, rice rats, and mice, and would strike at anything that moved, including the foot or a leg of an unsuspecting passerby. There were other potentially lethal entities lurking there as well: malaria-transmitting mosquitoes, and aquatic snails shedding their skin-penetrating stage of crippling blood fluke parasites. Either of these disease-causing agents could prove as deadly as the bite from any poisonous snake.

But not even these potential threats could deter Lian-Fu from focusing on the problem at hand. Frogs were scarcer now than when he was a child, he thought, as he resumed his hunt. One could easily trap ten to fifteen in an afternoon's safari when he was just a young apprentice, learning his craft at his father's knee. Lian-Fu's perception of the decreased density of frog populations was indeed correct. Globally, amphibians of all kinds are becoming rare. (More about this later in our story.) A few yards further, he startled a feral cat that had greater success at bagging its prey, having nearly finished eating its second frog of the evening. The remains were abandoned in haste as it stealthily disappeared into the protective embrace of twilight. How could he ever find what he had come for with all these other skilled hunters in action? The first stars of the evening shimmered on the watery surface of the paddy next to him, and he knew the day was ebbing into evening. Another uneventful half-hour passed. Lian-Fu now could barely see what lay just two feet in front of him. Deepening shades of blue blackness blurred the landscape. It was time to go. Then, almost as if on cue, he caught a glimpse of what he had come for. It was suspended in the water not more than a foot or two from where he stood, motionless. Its eyes peered out of its larger-than-normal body and shone in the alpenglow like two iridescent green marbles with jet-black centers. It was *the* frog.

Lian-Fu eyed his quarry with extreme caution, reckoning that this would be his one and only opportunity. He could not afford to miss. For all he knew, his son's life might hang in the balance, especially if the infection spread back along the contour of the eye, through the optic nerve and into the brain. Lian-Fu tried to remember all he had learned about catching frogs as a small child. He had recently acquired a sturdy throw net with drawstrings for that purpose, which he now slowly arranged with one hand behind his back. Lunging forward in a single motion, the huntsman threw it open and it enveloped the surprised animal, giving it no chance for escape. Lian-Fu pulled the drawstrings shut. Success! The oversized amphibian struggled mightily at first, flexing its powerful hind legs against the mesh of the trap, then fell into a motionless state of shock. The distraught father turned and raced for his village at full speed without so much as a brief consideration for his own safety that might abruptly end with a sudden strike from an innocent-looking piece of root that could transform in an instant into a surprised and aggressive viper. Nearing home, the rice farmer stopped to examine his catch one more time to make certain that it had blotches of olive markings on its body, and he marveled at its overall appearance, especially its girth. As a person who had been fond of collecting such animals when he was a child, he could not recall ever having seen one so out of proportion to his image of what a frog should look like.

At home, breathless and triumphant, he announced his presence to Su, who was in the family bedroom tending to their ailing son. Lian-Fu proceeded immediately with the preparation of his son's medication. He held the struggling animal down with one hand and, with the other firmly grasping a small kitchen knife, executed a precise, deep incision just in back of the skull, severing the spinal cord, finally putting the frog out of its misery. The grateful farmer uttered a brief prayer of thanks over the carcass, then, as instructed, carefully dissected out the thigh muscle and all the skin next to it. The still-quivering piece of muscle tissue was spread open and flattened to fit the contour of Gao's face. Lian-Fu draped the inner surface of the frog leg segment against the diseased area. The boy winced against its unfamiliar texture, its nervous reflexive trembling, and especially its coolness. But soon he adjusted to the presence of the poultice and in fact came to appreciate the soothing it afforded. A cloth bandanna secured the leg and skin flap in place, while the boy at last went into a deep sleep.

The drug was removed briefly for inspection of the wound over the next few days. Within two days the eye showed definite signs of improvement. By the end of the fourth day the swelling had gone down significantly, and Gao could see again for the first time since his accident. The redness and tenderness had also dissipated, along with the warmth induced by the intense inflammation caused by the pathological effects of the infection. The boy's family was overjoyed with relief, and they made a special trip to the shaman to thank him. Lian-Fu also slept well for the first time in more than a week. The next morning he woke up early and refreshed, returning to the rice fields with a light heart filled with the satisfaction of having persisted in his quest for the cure for his son's illness. A day later the second grader returned to class, much to the joy of his teacher and fellow students. All was well with the world, and there was reason to celebrate. Su prepared a rare family favorite for dinner, and they all gave thanks to nature for giving them a bountiful medicine chest of curatives.

But all was not what it appeared to be. Despite Gao's successful recovery, his life was to change for the worse yet a second time. The boy woke up two days after retuning to school with an uneasy feeling in his recovered eye. Something was not right. There was no pain this time, but the swelling had started again. The child feared his original infection had come back, but that was not the case. If not a bacterial infection, then what?

Before describing what happened next, let's take a moment to explain why the frog poultice worked. Because of the pioneering research of Dr. Michael Zasloff, we now know that frogs produce a potent antimicrobial peptide agent that Zasloff isolated and described. He named the curative compounds found in frog skin magainin.There is a whole family of related peptides found throughout the animal kingdom that are closely related to Zasloff's original compounds. They all serve a single purpose, namely, to protect fish, amphibians, and reptiles from their own assemblages of infectious agents, mostly bacteria, fungi, and single-celled parasites (protozoans). We also synthesize and secrete these compounds into our tears, saliva, and urogenital tracts. Magainins, chemically referred to as cationic peptides, have a molecular structure related to gramicidin, the world's first antibiotic isolated from a soil-dwelling microorganism. These related peptides all appear to work the same way, by literally punching holes through the protective outer membrane of the offending organism, killing it on the spot. But despite these encouraging findings, they are so potent that none of the ones isolated so far can be administered internally for use against our own pathogens.

Frogs produce as many as ten different varieties of magainins for each species. Since there are some 3,800 frog species alive in the world as of 2012, we would do well to investigate them to see which of their magainins work best against our own infections. This kind of survey is urgently needed in light of the fact that a great number of well-known antibiotics, including commonly used ones like penicillin, are not nearly as effective as they once were due to selection for microbial resistance.

Another reason to speed up the search for "user-friendly" magainins is that biologists around the world, within just the past ten years, have observed the crash of populations in a wide number of amphibian species. Some frog species have become extinct within that same time frame. The cause for the global decrease in amphibians appears to be a widely distributed fungal pathogen, *Batrachochytrium dendrobatidis*. Increased susceptibility to infection with that pathogen in a wide variety of amphibians may be linked to the reduction of ozone in our stratosphere caused by aerosolized chlorine compounds. These compounds find their way into the stratosphere and interfere with the formation of this protective layer that ordinarily absorbs UVB before it can cause us harm. It is a fact that over the past twenty years, increasing levels of UVB radiation have been measured in each succeeding year at Earth's surface. This damaging radiation may be what is inhibiting amphibian immune systems, analogous to the way that the HIV virus inhibits the human immune system, and with similar consequences.

In our fictional story, it was the magainin peptides from the frog skin that cured Gao's earlier bacterial infection. So what devastating event actually happened that eventually cost Gao the sight in his left eye? It was caused by a very unlucky side effect also introduced by the individual frog with which he was treated. Had Lian-Fu caught a different frog of the same species that night, things might have turned out for the better. Gao's second infection was due to the juvenile form of a tapeworm, *Spirometra mansonoides*. His new symptoms began with a tightness around the orbit of his left eye. It started to swell, but without the pain and tenderness of the first episode. In fact, the swelling was due to a growth hormone that the parasite secreted, which stimulated all cells receiving the signal to divide and expand their populations. The swelling progressed, becoming noticeable to others within the next few days. This medical condition is called sparganosis. *S. mansonoides* has a worldwide distribution, so sparganosis is a common disease in many regions. This finding implies that tens of thousands of humans rely on traditional medicines such as animal-derived poultices for curing everything from cuts and bruises to photophobia associated with chicken pox. In Gao's case, discomfort was evidenced by the boy's constant state of agitation and irritability. He began loosing sleep again, and so did everyone else in his family. Su and Lian-Fu were deeply saddened by this unanticipated consequence of the treatment. Worse yet, neither they nor their shaman could help any further. Gao was told by the old healer that he was most unfortunate, but he would have to live with his new disease, as many others had done before him.

When we think of a tapeworm, most of us conjure up a large, flat, ribbonlike, multisegmented organism (see chap. 6). Recall that the adult stage of all tapeworms lives in the lumen of the small intestine, regardless of the species of host it infects. Tapeworms also produce much smaller stages that live in other anatomical locations, and in a different set of animals. In the case of *S. mansonoides*, these intermediate hosts include fish, frogs, and snakes. Something that eats these cold-blooded hosts acquires the adult, and in the case of *S. mansonoides*, the cat or other feline is the final, or definitive, host.

When the boy received his poultice, the small, threadlike juvenile worm, lying just under the skin of the amphibian, detected the warmth of the child's face and migrated out of the frog skin, lodging under his lower eyelid. Parasitologists refer to this interesting activity as positive thermotactic behavior. In the laboratory this has been well characterized, and it is a common feature to all parasitic worms that infect warm-blooded animals. If anyone in Gao's family had had some knowledge of the tapeworm's life cycle, he or she could have easily removed the worm by simply exposing the inner surface of the lower lid and removing the small, whitish, threadlike object. The juvenile form of *S. mansonoides* rarely causes blindness in humans, but unfortunately in Gao's case the parasite did. He was particularly unlucky in that regard.

Hainan is home to some eight million people, and an astounding 30 percent of them have suffered at one time or another from this parasite-induced condition. That is a truly amazing number of cases. The incidence of the juvenile form of the tapeworm in frogs in Hainan must also be quite high to account for so many human infections. The prevalence of the adult tapeworm in wild and domestic cats has been recorded as high as 100 percent in some areas of that island.

So, in the end, every drug has its advantages and disadvantages. In Gao's case, the drug worked quite well, but its side effect was devastating. The side effect was not the usual kind that we expect from drugs. As illustrated in part 1 of this chapter, most of ours come to our attention in a box containing the drug of purchase with a folded piece of paper inside (the infamous package insert with all the warnings). In very small, nearly unreadable font, virtually every adverse reaction recorded during the extensive clinical trials required by law is revealed to the user. At least then we have the option not to proceed to take the drug if we think that the risks are too great compared with the stated benefits. Unfortunately, traditional medicines do not come with a set of instructions or a list of caveats as to what to expect if things go wrong. The less developed world is still obliged to make use of agents harvested directly from nature, and without the advantage of distilling out the active component from the rest of the material from which it was derived.

Plowshare Concepts

The pharmaceutical industries of the world have spent over sixty years identifying, developing, and marketing antibiotic compounds. Penicillin, streptomycin, and 136 other successfully exploited molecules were all derived from soil-dwelling microbes—fungi and actinomycetes. Despite the life-saving benefits that antibiotic therapies obviously have provided us, we are all too familiar with the plight we now face. Nearly all our commonly occurring bacterial infections, from syphilis and gonorrhea to tuberculosis and hospital-acquired staphylococcus, have taken the next step in their evolution and are now either partially or completely resistant to practically all commercially available antibiotics. If this trend continues, and there is no reason to think that it won't, soon we will have none at our disposal. But we still desperately need them, since the pathogenic microbes do not seem to have gone away just yet. Today drug development is expensive and time consuming. It takes nearly ten years before any newly discovered antibiotic makes it out of the laboratory and onto the pharmacist's shelf. The Food and Drug Administration's stringent set of regulations prohibits the release of any compound before first thoroughly testing it in animals and then ushering it through many expensive, time-consuming human clinical trials. The result is that most large drug companies have folded their tents and declared a truce with the microbes. It has proven too expensive to screen thousands of natural products that are generated randomly from fermentation flasks of exotic soil samples, only to have them flop in some assay or clinical trial. What's more, after a new drug does reach the market, resistance is an assured consequence of use. Then it is back to the drawing board after an all-too-brief respite from any given infectious disease. Other reasons for drug companies to give up research divisions dedicated to discovering new antibiotic compounds include declining revenues from drug sales owing to competition from generic smaller pharmaceutical companies, and the expiration of "cash cow" patents on older, more widely used antibiotics.

It is obvious to me that nature has all the answers, so what is our question? This is a dominant theme that I have revisited again and again throughout this book. If we are to hold off the pathogens at our doorstep, our question should be: where *will* our next generation of antimicrobial drugs come from? To answer this I suggest we first adopt a deeper, more respectful relationship with the natural world. We have to stop plowing over or cutting down entire ecosystems in favor of temporary monetary gains. Mining and farming interests in the rain forests come to mind as an example of such irresponsible behavior. We need time to explore more fully the natural world around us to allow us to identify a wider range of molecules of interest for our own use in fighting off infections. In learning how the rest of life arms itself against their own cadre of pathogens, we will come to realize that there are common grounds that unite us with all other nonparasitic animals and some plants too. Clearly the magainins fall into this general category, along with many other newly discovered molecules with antibiotic activity found in a broad swath of invertebrates (e.g., cecropins from the cecropia moth) and cold-blooded vertebrates (e.g., dermaseptins from the skin of certain species of frog). But despite all this new knowledge, it is hard to predict which family of defense peptides will eventually win the day.

In all natural systems, a balance is eventually achieved between any given host and its pathogens. It has been this way since complex ecosystems have been in existence. The same governing principles also applied to human populations prior to the advent of the antibiotic revolution. Are we going to have to experience that kind of thing each time a new, virulent strain of pathogenic microbe shows up at our door, now that most of these wonder drugs no longer work? Without antibiotics, our only defense, beside the development of vaccines, is an intact immune system, a highly functioning system of public health, and a bit of luck in avoiding exposure to any given infection in the first place.

So, in the end, I am convinced beyond any shadow of doubt that we need to remain on ready alert for the possibility of discovering new compounds from the natural world to feed into our diminishing arsenal of antimicrobials. In summary, I hope that what I have described in these pages will convince even the most skeptical as to the practical value of basic scientific knowledge. That knowing more about the lives of our pathogens, especially those we know little about today, will eventually improve our own lives. Spock said it best: "Live long and prosper." I believe that is everyone's wish. I also believe that our parasites, despite their own selfish behavior, can help us do just that.

GLOSSARY

antibodies a protein molecule (there are five varieties) produced by specialized white blood cells called B cells produced in response to exposure to foreign substances (e.g., bacterial toxins, structural components of parasites, viruses, fungi, bacteria). They can be proteins, complex sugars, toxins, etc., or active components of vaccines.

antigenic signature the complex of unique chemical components that, taken as a whole, help to characterize a parasite at the molecular level as identified by the immune system.

amino acid one of twenty chemical building blocks that make up all proteins. Each one has an acidic and a basic component to its chemical composition, allowing them to be connected into chains of peptides (see peptide).

capillary a small, one-red-cell in diameter (7 micrometers) tube that is part of the circulatory system. It is the most abundant of the blood vessels, and it is where the exchange of oxygen for carbon dioxide occurs in the tissues.

cytoplasm a general term for the inside of any cell, which typically contains the chromosome(s), and all other organelles that comprise that cell (e.g., mitochondrion, smooth and rough endoplasmic reticulum, nucleus, chloroplast).

dendritic cells primary immune cells whose job it is to process incoming foreign soluble antigens, preparing them for further processing by B cells, which then respond by making antibodies.

dermis the portion of the skin (the largest organ of our body) that lies just below the epidermis, the outermost covering of skin.

differentiated cell any cell with a specialized function (e.g., muscle, nerve, bone, white blood cell), as opposed to any cell with a generalized function (e.g., fibroblast, embryonic stem cell).

duodenum the first part of the small intestine, followed by the jejunum, then the ileum.

epidemiology the scientific study of how diseases behave in human populations.

eukaryote a cell that possesses a true nucleus (DNA enclosed inside a membrane).

filariae a generic term applied to a specific group of parasitic nematodes (see nematode), all of which are transmitted by arthropod vectors.

helminth a parasitic worm. Nematodes, trematodes, and cestodes fall into this general category.

host any organism harboring parasites.

intermediate host any organism harboring an intermediate stage of a parasite.

infective stage the developmental stage of a parasite capable of infecting its definitive host.

immunosuppressant any substance capable of inhibiting any or all portions of the immune system.

larva an immature stage of a nematode parasite.

lumen the hollow part of a tube.

magainin a family of related peptides produced by amphibians and reptiles with antimicrobial activity.

microvilli small, submicroscopic cellular projections that function to increase the surface area of the cell. Typically found on the cells lining the small and large intestines.

nematode a phylum of invertebrates, all of which are characterized by a smooth, nonsegmented outer cuticle, inside of which are nerves, muscles, an excretory system, a fully formed and functional gut tract, and reproductive organs. Typically the sexes are separate. The great majority of nematode species are free living and found throughout most terrestrial and many aquatic ecosystems. A smaller number are parasitic in a wide variety of host species, ranging from insects to humans.

parasite any organism that derives its shelter and general nutrition from another, larger organism.

pathogen any organism capable of inducing a state of ill health by the process of infection or the production of a toxin. Typically, representative pathogens are viruses, bacteria, fungi, protozoa, or helminthes.

pepsin a digestive enzyme (protein) synthesized in the stomach that, under acid conditions, helps in the digestion of ingested proteins.

peptide a small compound consisting of amino acids linked together by the amino and carboxy groups.

phage a special variety of virus that infects bacteria and other similar microbial life forms. Used in the science of molecular biology to transfer genetic material from one organism or cell to another.

proteomics the new scientific field of identifying the proteins encoded for by messenger RNAs.

protozoa single-celled organisms, all of with contain a nucleus. The vast majority are found in the environment, while a smaller number are parasitic on a wide variety of host species, ranging from insects to humans.

specialized cell see differentiated cell.

transmission zone an ecologically definable region in which the transmission of a given parasite is favored.

trematode a phylum of invertebrates, characterized by a surface tegument and a nonsegmented body type that consists of muscle, nerve, reproductive organs, and an excretory system. All are parasitic in a wide variety of host species, including most cold-blooded and warm-blooded vertebrates. The vast majority of them require intermediate hosts such as snails to complete their life cycles.

vector any organism that helps in the transmission of a parasite and that the parasite needs in which to grow into the infective stage for its definitive host. Typical vectors include mosquitoes (malaria), black flies (*Onchocera volvulus*), ticks (Lyme disease), fleas (plague), and mites.

venules small blood vessels that carry nonoxygenated blood from the capillaries to the veins. Veins carry blood back to the heart where it is pumped into the lungs for reoxygenation.

venus plexus the blood vessels (veins) that surround the urinary bladder.

villous tissue fingerlike projections of the small intestine that help to increase the overall absorptive surface area of the digestive tract.

FURTHER READING

Chapter 1

Despommier, Dickson. "*Trichinella spiralis*: The Worm That Would Be Virus."
Parasitology Today 6, 6 (1990): 193–96.
———. "How Does Trichinella Make Itself a Home?" *Parasitology Today*
(1998): 318–23.

Chapter 2

Ettling, John. *The Germ of Laziness: Rockefeller Philanthropy and Public Health
in the New South*. Cambridge: Harvard University Press, 1981.
Rockefeller Sanitary Commission for the Eradication of Hookworm Disease.
Records, 1909–1915. Sleepy Hollow, N.Y.: Rockefeller Archive Center.

Chapter 3

Babu, S., et al. "Cutting Edge: Diminished T Cell TLR Expression and Function
Modulates the Immune Response in Human Filarial Infection." *Journal of
Immunology* 176 (2006): 3885–89.
La Greca, F., and S. Magez. "Vaccination Against Trypanosomiasis: Can It Be
Done or Is the Trypanosome Truly the Ultimate Immune Destroyer and
Escape Artist?" *Human Vaccination* 11 (2011): 1225–33.

Rassi, A., Jr., and J. Marcondes de Reende. "American Trypanosomiasis (Chagas Disease)." *Infectious Disease Clinics of North America* 26 (2012): 275–91.

Skelly, P. J., and A. R. Wislon. "Making Sense of the Schistosome Surface." *Advances in Parasitology* 63 (2006): 185–284

Tran, M. H., et al. "Tetraspanins on the Surface of *Schistosoma Mansoni* Are Protective Antigens Against Schistosomaisis." *National Medicine* 12 (2006): 835–40.

Chapter 4

Boothroyd, J. "*Toxoplasma Gondii*: 25 Years and 25 Major Advances for the Field." *International Journal of Parasitology* 39 (2009): 935–46

Chapter 5

Hotez, P. J., et al. "Helminth Infections: The Great Neglected Tropical Diseases." *Journal of Clinical Medicine* (2008): 118.

Chapter 6

Desowitz, R. *New Guinea Tapeworms and Jewish Grandmothers*. New York: Norton, 1981.

Despommier, D. "Tapeworm Infections: The Long and the Short of It." *New England Journal of Medicine* 327 (1992): 727–28.

Shantz, P., et al. "Neurocysticercosis in an Orthodox Jewish Community in New York City." *New England Journal of Medicine* 327 (1992): 692–95.

Zimmer, C. "The Brain's Hidden Epidemic: Tapeworms Living Inside People's Brains." *Discover Magazine* (May 2012).

Chapter 7

Campbell, W. C., et al. "The Discovery of Ivermectins and Other Avermectins." In *Pesticide Synthesis Through Rational Approaches*. Edited by P. Magee, G. K. Kohn, and J. J. Menn, 5–50. ACS Symposium Series. American Chemical Society, 1984. http://pubs.acs.org/doi/abs/10.1021/bk-1984-0255.ch001.

Carter Center. *Guinea Worm Disease Eradication.* http://www.cartercenter.org/
health/guinea_worm/mini_site/index.html.
Freidlander, W. *The Golden Wand of Medicine.* Westport, Conn.: Greenwood
Press, 1992.

Chapter 8

Carter Center. *River Blindness Program.* http://www.cartercenter.org/health/
river_blindness/index.html.
Zasloff, M. "Megainins, a Class of Antimicrobial Peptides from Xenopus Skin:
Isolation and Characterization of Two Active Forms, and Partial cDNA
Sequence of a Precursor." *Proceedings of the National Academy of Sciences*
15 (1987): 5449–53.

INDEX

advances and benefits from parasites. *See* Plowshares concepts

The African Queen (movie), 58–59

African trypanosomes *(T.b. gambiense, T.b. rhodesiense):* about the long-lived nature, 2; determination of two distinct species, 45–47; early exploration and discovery, 44–45; human antibody response, 48–50; infective mechanisms, 47–48; life cycle, 50–51; treatments, 52

Alderman, Edwin A., 26

Alien, parasite depiction in the movies, xv

allergies and asthma, 39–40, 109–111

American trypanosome (*T. cruzi,* Chagas' disease): Chagas' disease, 53; discovery and naming, 52–53; infective mechanisms and survival, 2, 55–58; life cycle, 54–55

amino acid, 195

Ancylostoma caninum (dog hookworm), 31–32. *See also* hookworms

Ancylostoma celanicum, 31. *See also* hookworms

Ancylostoma duodenale, 25–26, 31, 35, 92. *See also* hookworms

anemia: iron deficiency, 27–38, 37; malaria, 26; "southern laziness," 25–28, 37; tapeworm and Vitamin B12, 136–37; whipworm infection, 102–103

antibiosis, principle of, 174

antibiotic resistance, 40–41, 177–79, 190–93

antibiotics: discovery of gramicidin, 174–75; discovery of penicillin, 175–76; discovery of streptomycin, 176–77; Plowshares concepts, 190–93

antibodies: defined, 195; immunity and response time, 47–49; specialized cell, 195

Chagas, Carlos, 52

Chagas' disease. *See* American trypanosome

Chain (Dr.), 175

children: ascaris infections, 93–95, 99, 102; congenital toxoplasma infection, 84; dracunculosis infection, 159–60; retardation from parasite infestation, 27, 38–39, 108; treatment for parasites, 28, 39–40, 101; trypanosome infection, 55; whipworm infection, 102–106

China Hookworm Commission (CHC), 29–31

cholera, 4, 25, 29, 91

Claxton, Philander Priestly, 26

Clemens, Samuel (aka Mark Twain), 27, 114

Cobbold, Thomas S., xiv

composting, 29

Cort, William Walter, 30

Crane, Stephen, 176

Crohn's disease, 39, 73, 109–111

Cruz, Ozwald, 52

Cryptosporidium parvum, 31, 75

Cuckler, Ashton (Dr.), 180

cytoplasm, defined, 195

Darwin, Charles, 7–8

DDT (dichlorodiphenyltrichloroethane, insecticide), 70, 166–67

Deliverance (movie), 27

dendritic cells, defined, 195

dermaseptins, antibiotic activity, 192

dermis, defined, 195

Desowitz, Robert (Dr.), xiv, 132

diabetes: breakthrough treatments, xvii; transplanting pig islet cells, 20–21; treatment derived from hookworms, 111; treatment derived from *Trichinella*, 21–22

diarrhea and dysentery: drinking water sources, 154, 158–61; group of organisms causing, 75–76; hookworms and sanitation, 29; trichuris infection, 102, 108

Dickens, Charles, 4

"dicksonized," defined, ix

differentiated cell, defined, 196

Diphyllobothrium latum (fish tapeworm): control and eradication, 139; host range and rate of infection, 135–36; infective mechanism and hosts, 115; infective sources and prevalence, 134–35; life cycle, 137; transmission and maturation, 136, 138–39; treatment, 139. *See also* tapeworm

Discovery Channel (TV documentaries), xiv–xv

DNA sequencing: applied to parasite research, xvii–xviii, 66, 71–72, 86; describing genes and proteins, 22; determining hookworm infection, 37; insight into immune function, 161–62; schistosomes, 66; testing for anemia, 37; *Toxoplasma gondii*, 86; trypanosomes, 47–48

Dracunculus medinensis (dracunculosis): biology and development, 156–58; control and eradication, 154, 156, 158–61, 162; geographic distribution and prevalence, 152, 154, 156; infective stage and